ERROR-FREE COMPUTATION

Why It Is Needed and Methods For Doing It

by
Robert Todd Gregory
University of Tennessee

ROBERT E. KRIEGER PUBLISHING COMPANY
HUNTINGTON, NEW YORK
1980

Original Edition 1980

Printed and Published by
ROBERT E. KRIEGER PUBLISHING COMPANY, INC.
645 NEW YORK AVENUE
HUNTINGTON, NEW YORK 11743

Printed in the United States of America

Library of Congress Cataloging in Publication Data

Gregory, Robert Todd, 1920-
Error-free computation.

Bibliography: p.
1. Approximation theory--Data processing. 2. Floating-point arithmetic. I. Title.
QA297.5.G73 511'.4'0285 80-23923
ISBN 0-89874-240-4

This volume is dedicated to

DAVID M. YOUNG

my esteemed colleague for many years

PREFACE

My purpose in writing this material is to acquaint mathematicians and computer scientists with some of the difficulties associated with attempts at approximating the arithmetic of the real field $(\mathbf{R}, +, \cdot)$ by using the (finite) set of floating-point numbers $\mathbf{F}$ in an automatic digital computer, and the effects of these attempts at approximating real arithmetic on the numerical solution of ill-conditioned problems.

In the first three chapters we discuss some of the differences between mathematics and numerical mathematics including the effects of inexact floating-point arithmetic in attempting to solve ill-conditioned problems. We also discuss numerically stable and numerically unstable computational algorithms and the scaling problem for large systems of linear algebraic equations.

In the last two chapters we introduce the results of recent attempts at eliminating the difficulties discussed in these earlier chapters by doing error-free arithmetic. Residue arithmetic (using a single modulus and also using several pairwise relatively prime moduli) is discussed in Chapter 4 and finite-segment p-adic arithmetic is discussed in Chapter 5.

The treatment is expository and important results are stated without proof, for the most part, with specific references given to detailed proofs elsewhere.

Special acknowledgements are due Ms. Cathy Henn and Mrs. Ann McArthur for the excellent job of typing the manuscript.

Knoxville, Tennessee
January, 1980

R. T. G.

CONTENTS

CHAPTER 1

MATHEMATICS vs NUMERICAL MATHEMATICS

I shall not attempt to give a formal definition of either mathematics or numerical mathematics. It is probably sufficient to repeat the frivolous (but surprisingly adequate) definition "mathematics is what mathematicians do, and numerical mathematics is what numerical mathematicians do".

In general, numerical mathematicians usually (but not always) are involved with automatic digital computers in some sense, when they do their numerical mathematics, whereas mathematicians are rarely involved with automatic digital computers when they do their mathematics. However, there are some mathematicians who use computers and there are some numerical mathematicians who never use computers. No claim is made that the two sets of people are disjoint.

1. The Real Number System

One way to introduce the subject under consideration is to examine the real number system. Mathematicians are quite familiar with the fact that the system $(\mathbf{R},+,\cdot)$, consisting of the set of real numbers $\mathbf{R}$ and the two binary operations addition and multiplication, constitutes a field. This means that we have the following properties for the elements in $\mathbf{R}$.

closure	$a + b = c$	$a\cdot b = d$	
commutativity	$a + b = b + a$	$a\cdot b = b\cdot a$	
associativity	$(a + b) + c = a + (b + c)$	$(a\cdot b)\cdot c = a\cdot(b\cdot c)$	
unique identities	$a + 0 = a$	$a\cdot 1 = a$	
unique inverses	$a + (-a) = 0$	$a\cdot a^{-1} = 1$	$a \neq 0$

right and left distributivity

$$a\cdot(b + c) = a\cdot b + a\cdot c$$
$$(a + b)\cdot c = a\cdot c + b\cdot c$$

The addition and multiplication operations, along with the existence of a unique additive inverse, $-a$, and a unique multiplicative inverse, a^{-1} , for each $a \in \mathbf{R}$ (except that a^{-1} is not defined for $a = 0$) allow us to define subtraction and division in $\mathbf{R}$. Of course, since a^{-1} is not defined for $a = 0$, division by zero is undefined. Thus, addition, subtrac multiplication, and division (when defined) may be performed without restriction and will never lead out of $\mathbf{R}$. We call this closed system the real number system.

Geometrically we exhibit $\mathbf{R}$ by placing the elements of $\mathbf{R}$ in one-to-one correspondence with the points on a line and we often refer to this line as the <u>real line</u>.

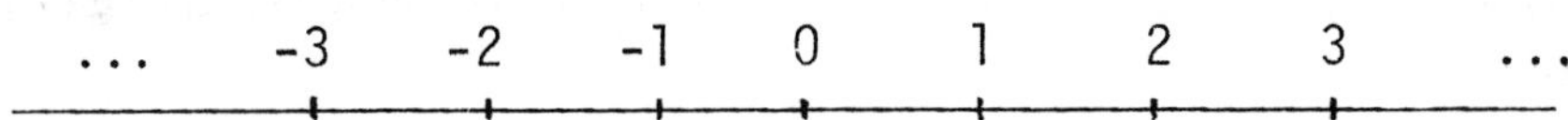

1.1 FIGURE. The Real Line.

Each point is associated with its distance from a fixed point called the ori those to the right are labeled positive and those to the left are labeled negative.

Each $x \in \mathbf{R}$ can be written in the form

$$x = m \cdot 2^e, \tag{1.2}$$

where e is an integer and m is a fraction satisfying the inequality

$$0 \leq |m| < 1. \tag{1.3}$$

We can always write

$$(1.4) \qquad |m| = b_1 \cdot 2^{-1} + b_2 \cdot 2^{-2} + \cdots + b_i \cdot 2^{-i} + \cdots$$

where the binary digits, b_i, lie in the range

$$(1.5) \qquad 0 \le b_i < 2.$$

If the expansion in (1.4) is periodic, x is a rational number; if it is not periodic, x is an irrational number.

It is usually the numerical mathematician rather than the mathematician who becomes involved in using the real number system for extensive computation and, in particular, he or she is usually interested in doing this computation using an automatic digital computer. However, this creates a problem because $\mathbf{R}$ cannot be represented inside a digital computer. The reason for this is obvious; $\mathbf{R}$ is an infinite set and a digital computer is a finite machine. A natural question then is "which real numbers are computer representable?"

2. Computer-Representable Numbers

Consider a fixed-word-length digital computer, designed for radix-two representation of numbers, with a word length of thirty-two bits (binary digits). Suppose twenty-four bits are reserved for the representation of $|m|$ in (1.4) and six bits are reserved for the representation of $|e|$. This leaves two bits for the purpose of designating the signs of m and e. Figure 2.1 demonstrates a possible distributation of the bits in a computer word.

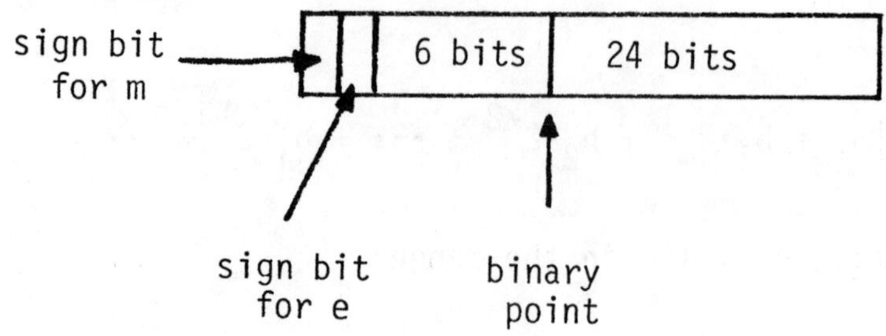

2.1 Figure. The location of the two sign bits, the six bits for $|e|$ and the twenty-four bits for $|m|$ in a computer word.

Assume that 0 represents a positive sign and 1 represents a negative sign in the first two positions* of the 32-bit configuration.

Consider the following example.

(2.2)
$$x = 5$$
$$= \frac{5}{8} \cdot 2^3$$

Thus, since m is positive,

(2.3)
$$m = |m|$$
$$= \frac{5}{8}$$
$$= 0.101000\ldots000000_{two}.$$

Likewise, since e is positive,

*On many computers a reversed sign bit convention is used for the sign of the exponent so that 1 represents a positive exponent and 0 represents a negative exponent. The reversed sign bit convention is never used for the sign of m , however.

(2.4)
$$\begin{aligned} e &= |e| \\ &= 3 \\ &= 11_{two}. \end{aligned}$$

Consequently, x = 5 has the following computer representation.

0	0	000011	101000...000000

2.5 Figure. The computer representation of $x = 5$.

Notice that we have chosen to pack the two sign bits, the six bits representing $|e|$, and the twenty-four bits representing $|m|$ into a 32-bit word in a specific way. Thus, it is easy to unpack the groups of bits and to reconstruct the number represented.

Consider the following example and the number it represents.

0	1	000100	100000...000000

2.6 Figure.

First, since the leftmost bit is 0, we observe that m is positive. Hence,

(2.7)
$$\begin{aligned} m &= |m| \\ &= 0.100000\ldots000000_{two} \\ &= \frac{1}{2}\ . \end{aligned}$$

Second, since the second bit is 1, we observe that e is negative. Hence

(2.8)
$$\begin{aligned} e &= -|e| \\ &= -\ 100_{two} \\ &= -\ 4. \end{aligned}$$

Consequently,

(2.9)
$$\begin{aligned} x &= m\cdot 2^{e} \\ &= \frac{1}{2}\cdot 2^{-4} \\ &= \frac{1}{32}\ . \end{aligned}$$

Obviously, in this computer, we can only represent real numbers for which

(2.10)
$$|m| = b_1\cdot 2^{-1} + b_2\cdot 2^{-2} + \ldots + b_{24}\cdot 2^{-24}$$

and

(2.11)
$$|e| = c_1\cdot 2^{0} + c_2\cdot 2^{1} + c_3\cdot 2^{2} + \ldots + c_6\cdot 2^{5}\ .$$

This automatically excludes the set of irrational numbers and all but a finite subset of the rational numbers. In other words, <u>the set of computer-representable numbers consists of a finite subset of the rationals</u>.

The set of computer-representable numbers is also known as the set of floating-point numbers since the binary point can be made to "float" merely by changing the exponent. This motivates us to use the symbol $\mathbb{F}$ for the set of computer-representable numbers.

Since $\mathbb{F}$ is finite it must contain a largest element and a smallest element. It is easily verified that $\mathbb{F}$ is symmetric about the origin and we need only examine the positive elements. In Figure 2.12 we see that the largest value of m and the largest value of e are presented. Hence, this represents the largest element in $\mathbb{F}$. Likewise Figure 2.13 exhibits the smallest positive element in F.

0	0	111111	111111...111111

2.12 Figure. The representation of the largest element in $\mathbb{F}$.

0	1	111111	000000...000001

2.13 Figure. The representation of the smallest positive element in $\mathbb{F}$.

In Figure 2.12 the values of m and e are, respectively,

(2.14)
$$\begin{aligned} m &= |m| \\ &= 0.111111\ldots111111_{two} \\ &= 1 - 2^{-24} , \end{aligned}$$

and

(2.15)
$$\begin{aligned} e &= |e| \\ &= 111111_{two} \\ &= 2^6 - 1 \\ &= 63 . \end{aligned}$$

Hence, the largest element in **F** is

(2.16)
$$\begin{aligned} x_{max} &= (1 - 2^{-24}) \cdot 2^{63} \\ &= 2^{63} - 2^{39} , \end{aligned}$$

which is approximately 2^{63}, in a relative sense. See Figure 2.21.

In Figure 2.13 the values of m and e are, respectively,

(2.17)
$$\begin{aligned} m &= |m| \\ &= 0.000000\ldots000001_{two} \\ &= 2^{-24} , \end{aligned}$$

and

(2.18)
$$e = -|e| = -111111_{two} = -63 \ .$$

Hence, the smallest positive element in **F** is

(2.19)
$$x_{min} = 2^{-24} \cdot 2^{-63} = 2^{-87} .$$

Remark. In the vicinity of x_{min} two adjacent computer-representable numbers are extremely close together. They are separated by exactly 2^{-87}. However, each time the value of e increases by unity the interval between two adjacent elements of **F** is doubled. Thus, the elements of **F** are not distributed evenly along the real line. (See Figure 2.21). In fact, in the vicinity of x_{max} the interval between two adjacent elements has grown to 2^{39}.

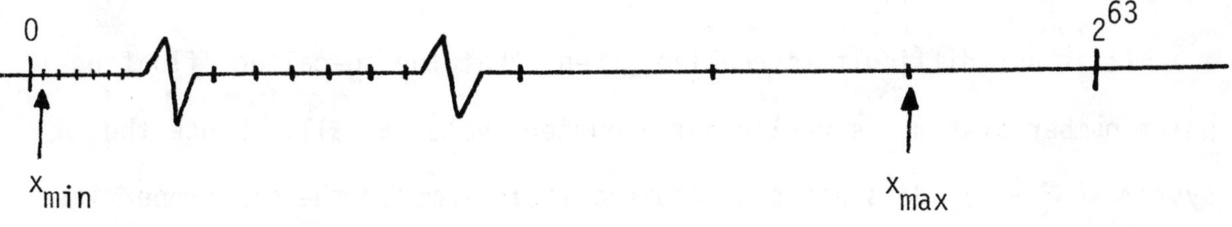

2.21 Figure. The intervals between adjacent elements in **F** (not drawn to scale).

If we summarize the observations above (and include additional observations) we have the following list of properties for the set of computer-representable numbers.

(a) $\mathbf{F}$ is a finite subset of the rational numbers.

(b) $\mathbf{F}$ is symmetric with respect to the origin and there are two representations of zero.

(c) The elements of $\mathbf{F}$ are not evenly distributed along the real line. The intervals between adjacent elements range from 2^{-87} near x_{min} to 2^{39} near x_{max}.

(d) Many "familiar" rational numbers are not elements of $\mathbf{F}$. For example, such numbers as $\frac{1}{3}$, $\frac{5}{6}$, and $\frac{1}{10}$ are excluded since the only candidates for membership in $\mathbf{F}$ are rational numbers of the form $\frac{p}{q}$, where q is a power of two.

(e) The system $(\mathbf{F},+,\cdot)$ does not constitute a field.

It is not difficult to realize, then, that the so-called "floating-point number system" is really not a number system at all. Since the system $(\mathbf{F},+,\cdot)$ does not constitute a field some of the six properties of a field (listed in Section 1) are no longer valid for all elements in $\mathbf{F}$. We cannot perform addition, subtraction, multiplication and division (when defined) without being lead out of $\mathbf{F}$ in most instances.

For example,

$$(2.22)\qquad \begin{cases} x_1 = \frac{1}{2}\cdot x_{min} \\ x_2 = 2\cdot x_{max} \\ x_3 = \frac{x_{max}}{2} - 64 \\ x_4 = 2^{14} + 2^{-14} \end{cases}$$

are four rational numbers not in $\mathbf{F}$ and yet each is computed using elements of $\mathbf{F}$.

3. Floating - Point Arithmetic

A mathematician is completely frustrated by $\mathbf{F}$ because of the five properties listed in the previous section (in particular, by the lack of the closure property). On the other hand, a numerical mathematician realizes that he must control his frustration and accept the challenge of trying to simulate the arithmetic of the real number system by using the floating-point arithmetic of a computer. Consequently, he often (but not often enough) aids the engineers in designing the electronic circuits for addition, subtraction, multiplication and division in order to achieve a simulation of the real number system that is useful.

In Figure 2.21 we exhibit that portion of the real line between the origin and 2^{63}. Elements of $\mathbf{F}$ are shown as "tick marks" on the line. Obviously, if we wish to represent a real number x in the range

(3.1) $$0 \leq x < 2^{63},$$

we select the element $\hat{x} \in \mathbf{F}$ which is closest to x. (A special rule is used for any real number which lies at the midpoint of the interval between two adjacent elements of $\mathbf{F}$.)

Consequently, in designing floating-point arithmetic it is customary to "force" closure in such a way that if $x \in \mathbf{R}$ is the exact result of an arithmetic operation, the result of the corresponding floating-point operation is defined to be $\hat{x}$, the element in $\mathbf{F}$ which is closest to x.

In order to introduce notation, let

(3.2) $$x = a + b$$

be the exact result of the addition of two elements of $\mathbf{F}$ and let

(3.3) $$\hat{x} = fl(a + b)$$

be the result using floating-point addition.

Floating-point arithmetic obviously is not exact, in general, and we call the difference

(3.4) $$\begin{aligned} \varepsilon &= \hat{x} - x \\ &= fl(a + b) - (a + b) \end{aligned}$$

the <u>rounding error</u> in forming the sum. We have similar expressions for the other arithmetic operations.

It is not just the lack of closure in floating-point operations which causes floating-point arithmetic to differ from the arithmetic of the real number system. There is also a lack of associativity, and distributivity caused by the rounding errors. For example, the four numbers

$$x_1 = fl\left[\frac{a\cdot(b+c)}{d}\right] \tag{3.5}$$

$$x_2 = fl\left[\frac{a}{d}\cdot(b+c)\right] \tag{3.6}$$

$$x_3 = fl\left[\frac{a\cdot b + a\cdot c}{d}\right] \tag{3.7}$$

and

$$x_4 = fl\left[\frac{c+b}{d}\cdot a\right] \tag{3.8}$$

quite probably are different elements of $\mathbf{F}$ even though the four expressions in the brackets are algebraically equivalent.

For these four numbers to be equal, the rounding errors in the floating-point operations would have to be such that, for example,

$$fl[a\cdot fl(b+c)] = fl[fl(a\cdot b) + fl(a\cdot c)] \ , \tag{3.9}$$

for all choices of a,b, and c. There is no computer known to this writer for which (3.9) holds in every case.

4. The Analysis of Errors

Whereas a mathematician usually works with the real number system and assumes that arithmetic is always exact, a numerical mathematician usually works with the "floating-point number system" of some computer and assumes that arithmetic is (almost) never exact. Consequently, the analysis of errors is a major activity of most numerical mathematicians.

Soon after the advent of the automatic digital computer in the 1940's several distinguished mathematicians, among them John von Neumann, found themselves becoming numerical mathematicians by virtue of the fact that they were involved in formulating the scientific problems whose solutions would be attacked by some of the first automatic digital computers. They were also involved in aiding the engineers in the logical design of some of the earliest computer circuits. Consequently, von Neumann made major contributions to the development of this newly emerging field. The first computers used the radix-ten representation of numbers but von Neumann lead the way in switching to the radix-two representation. (Almost all of today's computers use a radix which is two or a power of two.)

It was obvious to von Neumann and his colleagues that the numerical solution of a large system of linear algebraic equations was high on the priority list of problems for which an automatic digital computer would be absolutely essential in computing a solution. Since they were completely aware of the differences between the arithmetic of real numbers and the inexact arithmetic of computers* they became concerned about the

*It must be pointed out, however, that their crude computers used fixed-point arithmetic rather than floating-point arithmetic since floating-point hardware was not available at the time.

accumulation of errors which would result when a large system of linear algebraic equations might be solved using an automatic digital computer. They did a thorough analysis of the accumulation of errors, based on a numerical solution using Gaussian elimination, and the results were published by von Neumann and Goldstine [1947] in a paper which has become a classic. Many researchers today date modern numerical mathematics from 1947, the year of the publication of that first detailed error analysis.

Another problem high on the priority list for solution using a digital computer was the algebraic eigenvalue - eigenvector problem. A detailed error analysis of an algorithm for finding the eigenvalues of a real symmetric matrix was published as an Oak Ridge National Laboratory Report by Wallace Givens [1954].

The von Neumann error analysis is usually referred to as a <u>forward</u> error analysis in which one keeps track of the errors as they are introduced and develops bounds on the accumulated errors. The Givens error analysis, on the other hand, is an early example of a <u>backward</u> error analysis in which one takes the point of view that the computed solution to the problem must be the exact solution of a perturbed problem. This enables us to replace our concern over rounding errors introduced during the computation by a concern over the effect of introducing perturbations in the original data.

To illustrate backward error analysis, suppose we are solving the set of linear algebraic equations

(4.1) $$Ax = b,$$

with $A \in \mathbf{R}^{nn}$ and $b \in \mathbf{R}^{n}$, and suppose we obtain the computed solution $\hat{x}$. Let h be the error vector

(4.2) $$h = \hat{x} - x.$$

Suppose we take the point of view that $\hat{x} = x + h$ is the exact solution of a perturbation of (4.1) where the perturbation is in b . Then we may write

(4.3) $$A(x + h) = b + k,$$

where k is the perturbation vector.

Wilkinson [1963, p. 91] gives the following bound* on the relative error in x as a function of the relative error in b .

(4.4) $$\frac{\|h\|}{\|x\|} \leq N \cdot \frac{\|k\|}{\|b\|}$$

where

(4.5) $$N = \|A\| \cdot \|A^{-1}\|$$

Thus, instead of accumulating a bound on the errors as we proceed through the computation (forward error analysis) we have a bound on the relative error in the solution in terms of the relative size of the perturbation vector k .

Consider the numerical example (see Forsythe and Moler [1967], p. 24) in which

---

*We assume that the vector and matrix norms used in this treatise are consistent, that is, $\|Ax\| \leq \|A\| \cdot \|x\|$ for every matrix A and every vector x.

$$(4.6) \qquad A = \begin{bmatrix} 1 & 0.99 \\ 0.99 & 0.98 \end{bmatrix}, \qquad b = \begin{bmatrix} 1.99 \\ 1.97 \end{bmatrix},$$

and assume that a computed solution is

$$(4.7) \qquad \hat{x} = \begin{bmatrix} 3 \\ -1.0203 \end{bmatrix}.$$

If we substitute this solution into the original equations, that is if we form $A\hat{x}$, we obtain $b + k$, where

$$(4.8) \qquad k = \begin{bmatrix} -0.000047 \\ 0.000106 \end{bmatrix}.$$

Thus, $\hat{x}$ is the exact solution to the perturbed system (4.3) with k given by (4.8).

Since $\|k\|$ is small, we might be tempted to assume that $\hat{x}$ is a good approximation to x and that $\|h\|$ is also small. However, in comparing the relative values of these norms we see, from (4.4), that the factor N must be taken into account. It turns out, in this example, that

$$(4.9) \qquad x = \begin{bmatrix} 1 \\ 1 \end{bmatrix}$$

is the exact solution (by inspection) which implies that the error vector is

$$(4.10) \qquad h = \begin{bmatrix} 2.0000 \\ -2.0203 \end{bmatrix}.$$

Obviously, $\|h\|$ is not small relative to $\|k\|$. In fact, if we compare $\|h\|/\|x\|$ with $\|k\|/\|b\|$ we observe that the first quantity is greater than the second by about four orders of magnitude!

The only conclusion we can come to, in view of the fact that (4.4) must hold, is that the constant N in that inequality must be very large. It turns out that this is indeed the case and

$$N \doteq 39,600. \tag{4.11}$$

We shall discuss this example again in Chapter 2 and assign a name to the constant N.

5. The Analysis of Computational Algorithms

Another activity which seems to be central in the life of a numerical mathematician is his involvement in the evaluation of competing computational algorithms for solving a given problem, for example the problem in (4.1), and his involvement in the discovery of new and better algorithms for solving the problem. It should be pointed out that part of the evaluation of an algorithm is an error analysis referred to in the previous section.

As an illustration of the need for algorithm evaluation we recall from our high school algebra course that Cramer's Rule provides us with an explicit algorithm for solving the set of linear algebraic equations $Ax = b$ in (4.1). If

$$d = \det A, \tag{5.1}$$

and if

$$d_i = \det A_i, \tag{5.2}$$

where A_i is the matrix obtained by replacing the ith column of A by b, then the components of the solution vector x are

$$x_i = \frac{d_i}{d}, \qquad i = 1,2,\ldots,n. \tag{5.3}$$

This is an elegant result in the eyes of the mathematician who probably takes the point of view that this problem has a "closed solution" and needs no further examination. Nothing could be farther from the truth, from the numerical mathematician's point of view, because he usually finds Cramer's Rule completely useless for systems of linear algebraic equations with $n > 2$.

To evaluate each of the determinants $d, d_1, d_2, \ldots, d_n$, required in (5.3) by expanding in terms of minors, we must use $p_n n!$ multiplications, where p_n satisfies the inequality *

$$1 \leq p_n < (e-1) , \tag{5.4}$$

and p_n approaches $(e-1)$ as n becomes large. Thus, if $n = 20$, we are faced with the task of performing on the order of 21! multiplications in order to evaluate the twenty-one determinants $d, d_1, d_2, \ldots, d_{20}$, in terms of minors. This task would require a modern digital computer more than a million years of continuous operation!

Even if the sophisticated methods of Chiò (see Kunz [1957], p. 217) are used for the determinant evaluation, the fantastic reduction in the number of operations is not sufficient to make Cramer's Rule competitive with Gaussian elimination.

---

* $e = 2.71828\ldots$

It is not necessary to use a major computational problem such as the linear equations problem to illustrate the need for algorithm analysis and evaluation. Consider, for example, the "simple" problem of finding the two roots of a quadratic equation with real coefficients, that is, the roots of

$$ax^2 + bx + c = 0. \tag{5.5}$$

There is an excellent article by Forsythe [1969] which explores in great detail the difficulties associated with the accurate computation of the two roots, using floating-point arithmetic, when a,b, and c are allowed to be arbitrary floating-point numbers.

He makes it clear, for example, that when $abc \neq 0$ it is naive to use both of the formulas

$$x_1 = \frac{-b + \sqrt{b^2 - 4ac}}{2a} , \tag{5.6}$$

and

$$x_2 = \frac{-b - \sqrt{b^2 - 4ac}}{2a} , \tag{5.7}$$

which we learned in a high school algebra course because, when $4ac$ is small relative to b^2, there will be severe cancellation (due to the subtraction of two equal or nearly equal numbers) in one of the two formulas, with a subsequent loss of significant digits.

To illustrate this fact, Forsythe considers the solution of the quadratic equation

$$x^2 - 100{,}000x + 1 = 0 \tag{5.8}$$

using eight-decimal-digit floating-point arithmetic. Since

$$a = 0.10000000 \times 10$$
$$b = -0.10000000 \times 10^6$$
$$c = 0.10000000 \times 10$$

we have

$$fl(b^2) = 0.10000000 \times 10^{11}$$
$$fl(4a) = 0.40000000 \times 10$$
$$fl(4ac) = 0.40000000 \times 10$$
$$fl(b^2 - 4ac) = 0.10000000 \times 10^{11}$$
$$fl\sqrt{b^2 - 4ac} = 0.10000000 \times 10^6$$
$$fl\left(-b - \sqrt{b^2 - 4ac}\right) = \text{zero} \qquad \text{cancellation!}$$

and so (5.7) yields the answer

$$\hat{x}_2 = 0 \ . \tag{5.9}$$

The value of x_2, correct to eleven significant digits, is actually

$$x_2 = 0.000010000000001 \tag{5.10}$$

so that $\hat{x}_2$ has a relative error of about one hundred percent. If (5.6) is used, there is no cancellation and the computed value of x_1 is

$$\hat{x}_1 = 0.10000000 \times 10^6 \ . \tag{5.11}$$

Since the value of x_1, correct to eleven significant digits, is

$$x_1 = 99999.999990 , \tag{5.12}$$

$\hat{x}_1$ is correct to eight significant digits and this is all we can hope for in an eight-digit floating-point calculation.

Notice that cancellation is possible in (5.6) only if $b > 0$ and in (5.7) only if $b < 0$. Consequently, we should use only the appropriate one of the pair (the formula with no cancellation) to compute the first root and the second root should be computed using the fact that

$$x_1 \, x_2 = \frac{c}{a} . \tag{5.13}$$

See Chapter 2, Section 2, for additional comments about this problem.

We have mentioned only one of the difficulties discussed by Forsythe in his article on quadratic equations. He mentions several other, less obvious, difficulties which we shall not go into here, for lack of space. In his concluding remarks, Forsythe states that "The quadratic equation is one of the simplest mathematical entities and is solved almost everywhere in applied mathematics. Its actual use on a computer might be expected to be one of the best understood of computer algorithms. Indeed it is not $\cdots$. The fact that the obvious algorithm is so subject to rounding error is not very widely known by computer users $\cdots$. Thus, even in this elementary problem, we are working at the frontiers of common computing knowledge."

5.14 Remark It is not surprising that complex algorithms require extensive analysis before they can be implemented as working computer programs, in view of the experience of Forsythe in analysing one of the simplest of algorithms. It may surprise some, however, to realize that the type of digital computer, the computer's operating system, and the programming language in which an algorithm is coded often play major roles in the computer implementation of an algorithm. (It is a long step from the mathematical description of an algorithm to its efficient implementation on a digital computer). Thus, a numerical mathematician is often a computer scientist, as well.

CHAPTER 2

ILL-CONDITIONED PROBLEMS vs NUMERICALLY UNSTABLE ALGORITHMS

In the last chapter a great deal of emphasis was placed on the fact that a numerical mathematician spends much of his time discovering new computational algorithms and in analysing and implementing computational algorithms on a digital computer. In this chapter we want to differentiate between certain difficulties which may be encountered because of a computational algorithm used to solve a problem and certain difficulties which may be encountered because of the problem itself, independent of the algorithm.

1. Ill-Conditioned Problems

The solution to a problem in mathematics is a function of the data which describe the problem. For example, the determinant of a real square matrix

(1.1) $$A = (a_{ij})$$

is a continuous function of the matrix elements, a_{ij} , and the roots of the polynomial equation (with real coefficients)

(1.2) $$a_0 + a_1x + \dots + a_sx^s = 0$$

are continuous functions of the coefficients $a_0, a_1, \dots, a_s$.

Many numerical problems, therefore, can be described mathematically by a mapping of the form

(1.3) $$f : \mathbf{D} \subset \mathbf{R}^p \longrightarrow \mathbf{R}^q.$$

The p components of a vector d in $\mathbf{D}$ are the data that determine the problem and the q components of the vector f(d) constitute the solution to the problem. In the first example above, if $A \in \mathbf{R}^{nn}$, then $p = n^2$ and $q = 1$. In the second example, if all roots are real, p=s+1 and q=s. We can call $\mathbf{D} \subset \mathbf{R}^p$ the "data space" and $\mathbf{R}^q$ the "solution space".

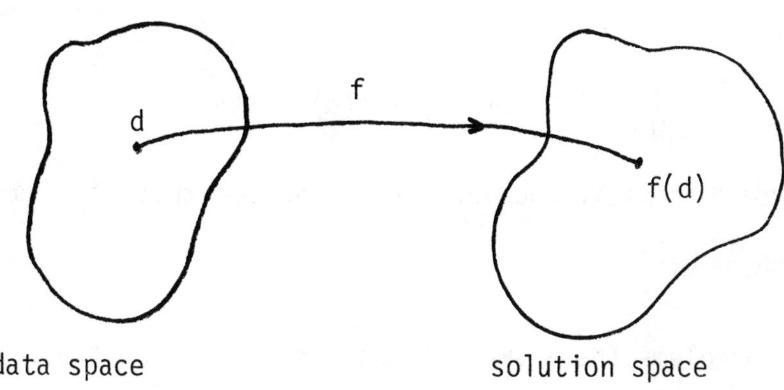

1.4 Figure. The mapping $f : \mathbf{D} \subset \mathbf{R}^p \longrightarrow \mathbf{R}^q$.

An interesting mathematical question to ask about any problem is how sensitive is the solution f(d) to small perturbations in d ? This question has practical implications because, in many problems, d may not be known exactly. If the data are obtained from physical measurements, we know that they are in error because, except for counting (in which case the results are positive integers), all measurements involve some error.

Suppose we do not know d but we do know $d + \delta$, where δ is a small perturbation in d . The sensitivity question now becomes how much does $f(d + \delta)$ differ from $f(d)$? To put it in more mathematical terms, suppose $\|\cdot\|$ is a vector norm and suppose

$$\|\delta\| < t_1 \tag{1.5}$$

and

$$\|f(d + \delta) - f(d)\| < t_2. \tag{1.6}$$

If t_1 is a small positive number, can we assume that t_2 is also a small positive number?

For many problems the answer is yes, in which case the solution is not sensitive to small perturbations in the data. Such a problem is called <u>well conditioned</u>. However, for some problems the answer is no, in which case the solution is sensitive to small perturbations in the data. This case is illustrated in Figure 1.7. Such a problem is called <u>ill conditioned</u>.

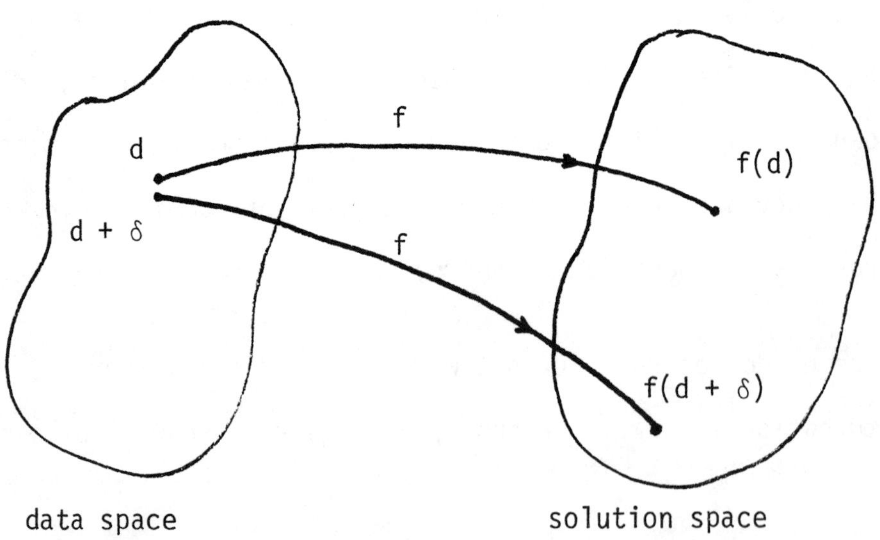

1.7 Figure. An ill-conditioned problem.

As a numerical example of an ill-conditioned problem (see Gregory and Karney [1978], p. 50) consider the two matrices

(1.8)
$$A = \begin{bmatrix} -73 & 78 & 24 \\ 92 & 66 & 25 \\ -80 & 37 & 10 \end{bmatrix}$$

and $A + E$, where

(1.9)
$$E = \begin{bmatrix} 0 & 0 & 0 \\ 0 & 0 & 0 \\ 0 & 0 & 10^{-2} \end{bmatrix}.$$

That the determinant of A is extremely sensitive to small perturbations in the elements is shown by the fact that

(1.10)
$$\det A = 1 ,$$

whereas

(1.11)
$$\det (A + E) = -118.94 ,$$

and these results are <u>exact</u>.

Thus, the fact that A is ill-conditioned with respect to the determinant problem is a <u>mathematical property</u> of A and it has absolutely nothing to do with the numerical algorithm used to compute the determinant. Incidently, if A is ill-conditioned, with respect to the determinant problem so is $A + E$ because either matrix may be considered to be the result of perturbing the other.

To exhibit a second example we return to the numerical example in Section 4 of Chapter 1 described by (4.6) - (4.10). This time we interpret things differently, however. Our point of view here is that we have two linear systems

$$(1.12) \qquad \begin{bmatrix} 1 & 0.99 \\ 0.99 & 0.98 \end{bmatrix} \begin{bmatrix} x_1 \\ x_2 \end{bmatrix} = \begin{bmatrix} 1.99 \\ 1.97 \end{bmatrix}$$

and

$$(1.13) \qquad \begin{bmatrix} 1 & 0.99 \\ 0.99 & 0.98 \end{bmatrix} \begin{bmatrix} x_1 \\ x_2 \end{bmatrix} = \begin{bmatrix} 1.989903 \\ 1.970106 \end{bmatrix}$$

whose points in "data space" (see Figure 1.7)

$$(1.14) \qquad d = \begin{bmatrix} 1 \\ 0.99 \\ 0.99 \\ 0.98 \\ 1.99 \\ 1.97 \end{bmatrix}, \qquad d + \delta = \begin{bmatrix} 1 \\ 0.99 \\ 0.99 \\ 0.98 \\ 1.989903 \\ 1.970106 \end{bmatrix}$$

are close together and whose points in "solution space"

$$(1.15) \qquad f(d) = \begin{bmatrix} 1 \\ 1 \end{bmatrix}, \qquad f(d + \delta) = \begin{bmatrix} 3 \\ -1.0203 \end{bmatrix}$$

are not close together. Thus, again we have a problem whose solution is sensitive to small perturbations in the data (an ill-conditioned problem).

In the first example, a perturbation of 10^{-2} in the element a_{33} changed the determinant from 1 to -118.94. In the second example, the perturbations -0.000047 and 0.000106 in the components of the vector on the right side of (1.12) changed the solution from $x_1 = 1$, $x_2 = 1$ to $x_1 = 3$, $x_2 = -1.0203$.

In Chapter 1 we used the notation x for $f(d)$ and $\hat{x}$ for $f(d+\delta)$, where $\hat{x}$ was assumed to be an approximation to the true solution x . It turns out that the vector k (see (4.8) in Chapter 1), in addition to its interpretation as the perturbation vector in (4.3), can be interpreted as the residual vector

$$k = A\hat{x} - b . \qquad (1.16)$$

Thus, in this example, we see that k is a small residual vector and yet $\hat{x}$ is not a good approximation to x . This illustrates a well-known fact that, for ill-conditioned problems, a poor approximate solution $\hat{x}$ can give a small residual vector k . Consequently, the size of the residual vector is not always a good test of how good a computed solution may be.

It is the inequality (4.4) with N defined in (4.5) which provides the clue for understanding what is happening here. If N is small, then a small perturbation vector k guarantees that h is small and so $\hat{x}$ is a good approximation to the true solution x . However, if N is large, we cannot be certain. Certainly if $\hat{x}$ is a poor approximation to x , that is, if h is large, then N will have to be large; but if N is large, it does not necessarily follow that h is large and that $\hat{x}$ is a poor approximation to x .

Wilkinson [1963, p. 91] defines $N = \|A\| \cdot \|A^{-1}\|$ as a condition number for the linear equations problem $Ax = b$. It is a constant* whose smallness tells us how well conditioned our problem may be.

Incidentally, since

$$(1.17) \qquad \begin{aligned} 1 &= \|I\| \\ &= \|A \cdot A^{-1}\| \\ &\leq \|A\| \cdot \|A^{-1}\| \\ &= N \end{aligned}$$

we know that the condition number is never less than unity.

1.18 REMARK In this section we have mentioned a condition number for the problem $Ax = b$. This is not a condition number for other computational problems involving A , however. It turns out that it is necessary to do an independent analysis of each problem in order to discover a condition number for that problem.

---

* The condition number, of course, is a function of the particular matrix norm used in defining N.

2. Numerically Unstable Algorithms

When we select an algorithm for solving a particular problem (described by the vector d) and when we implement this algorithm on a digital computer, we are, in effect, defining a function F associated with f in (1.3). Our objective is to have $F(d)$ be a reasonable approximation to the exact solution $f(d)$. However, this is not always possible, as we shall see.

Stewart [1973, pp. 76-77] makes the point that if d contains the data describing an ill-conditioned problem and if we begin with perturbed data, $d + \delta$, then no matter how good F is, we will get a poor approximation to the solution $f(d)$. He also points out that, in choosing F, we certainly cannot demand that F produce a solution to an ill-conditioned problem "more accurately than the data warrents". On the other hand, F should not introduce "larger inaccuracies of its own".

Stewart goes on to suggest that F should be such that for every $d \in \mathbf{D}$, there exists a nearby point $d + \delta$ such that $f(d + \delta)$ is close to $F(d)$. To state this another way, F should be such that $F(d)$ is always close to "the exact solution of a slightly perturbed problem". If the algorithm F has this property, it is called <u>numerically stable</u>. If it does not, it is called <u>numerically unstable</u>.

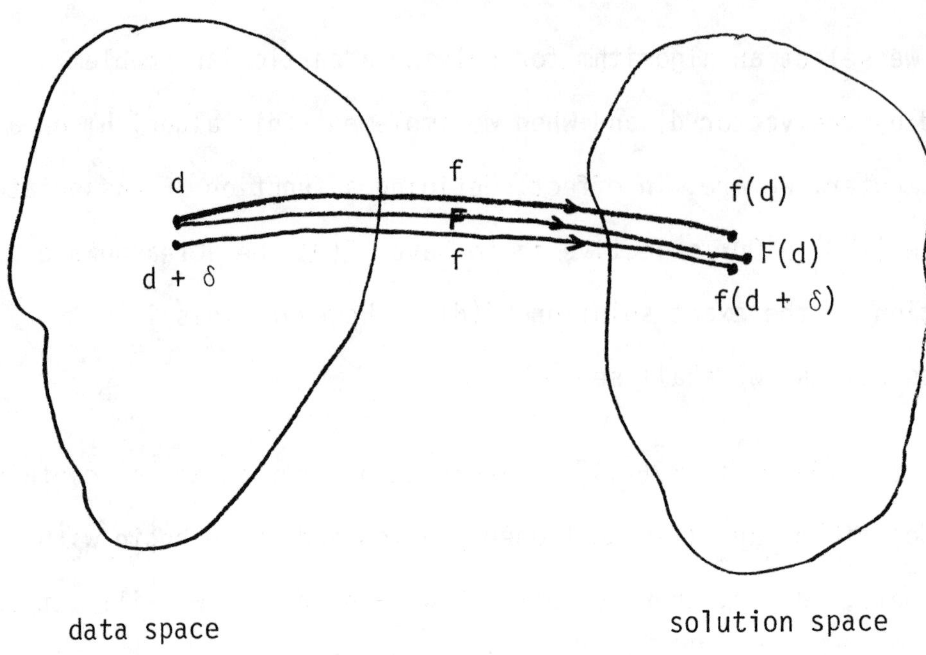

2.1 Figure. A numerically stable algorithm applied to a well-conditioned problem.

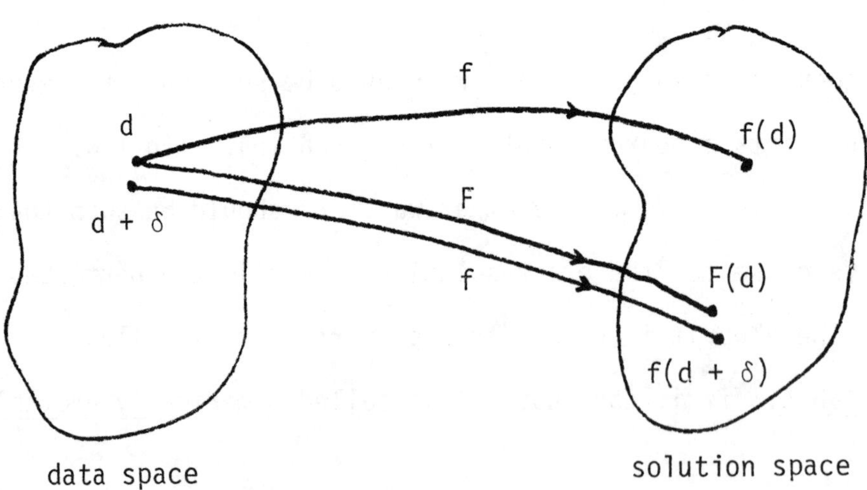

2.2 Figure. A numerically stable algorithm applied to an ill-conditioned problem.

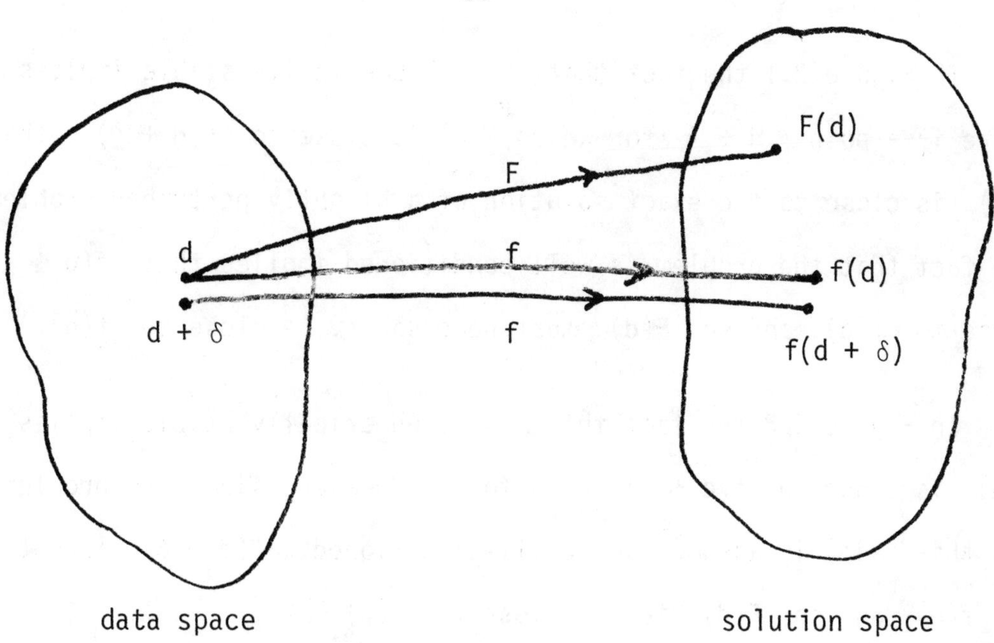

2.3 Figure. A numerically unstable algorithm applied to a well-conditioned problem.

In Figure 2.1 the fact that F is numerically stable implies that there is a point $d + \delta$ for which F(d) is close to $f(d + \delta)$, that is, F(d) is close to the exact solution of a slightly perturbed problem. The fact that the problem is well-conditioned implies that $f(d + \delta)$ is close to f(d) and so F(d) must necessarily be close to f(d).

In Figure 2.2 the fact that F is numerically stable implies that F(d) is close to $f(d + \delta)$ as before. However, since the problem (in this case) is assumed to be ill-conditioned, $f(d + \delta)$ is not close to f(d) and so F(d) is not close to f(d).

Now we illustrate the case of a numerically unstable algorithm applied to a well-conditioned problem. Recall (5.8) - (5.13) in Chapter 1 where we exhibited Forsythe's example of a quadratic equation and his unwise choice of an algorithm for computing the second root. By using (5.7) he permitted cancellation to occur due to the subtraction of two equal quantities (they were only equal to eight significant digits, however) with the resulting loss in accuracy indicated by the fact that the computed answer had a relative error of about one hundred percent. The formula in (5.7) was chosen deliberately to illustrate a numerically unstable algorithm.

We indicated that cancellation could be avoided by computing the first root using (5.6) and the second root using (5.13). Another choice for computing the second root would be to replace (5.7) by the algebraically equivalent expression

(2.4) $$x_2 = \frac{2c}{-b + \sqrt{b^2-4ac}} \quad ,$$

obtained from (5.7) by rationalizing the numerator. In this way we can change the algorithm to one which is numerically stable.

As a second example of a numerically unstable algorithm applied to a well-conditioned problem Wilkinson* [1965, p. 216] displays the following system of linear algebraic equations, $Ax = b$, and its solution in six-decimal-digit floating-point arithmetic.

(2.5) $$\begin{bmatrix} 0.000003 & 0.213472 & 0.332147 \\ 0.215512 & 0.375623 & 0.476625 \\ 0.173257 & 0.663257 & 0.625675 \end{bmatrix} \begin{bmatrix} x_1 \\ x_2 \\ x_3 \end{bmatrix} = \begin{bmatrix} 0.235262 \\ 0.127653 \\ 0.285321 \end{bmatrix} .$$

In order to perform the first step of Gaussian elimination, we choose a_{11} as the <u>pivot</u> and form the two multipliers

(2.6) $$m_{21} = -\frac{a_{21}}{a_{11}} ,$$

and

(2.7) $$m_{31} = -\frac{a_{31}}{a_{11}} .$$

If we multiply the first (or pivotal) equation by m_{21} and add it to the second equation, we eliminate x_1 from the second equation. Likewise, if we multiply the first equation by m_{31}, and add it to the third

---

*See also Young and Gregory [1973], p. 804.

equation, we eliminate x_1 from the third equation. This produces the reduced system of equations

$$\begin{bmatrix} 0.000003 & 0.213472 & 0.332147 \\ 0 & -15334.9 & -23860.0 \\ 0 & -12327.8 & -19181.7 \end{bmatrix} \begin{bmatrix} x_1 \\ x_2 \\ x_3 \end{bmatrix} \doteq \begin{bmatrix} 0.235262 \\ -16900.5 \\ -13586.6 \end{bmatrix} . \tag{2.8}$$

For the second step of Gaussian elimination we choose* a_{22} as the pivot and form the multiplier

$$m_{32} = -\frac{a_{32}}{a_{22}} . \tag{2.9}$$

Now, if we multiply the second (or pivotal) equation by m_{32} and add it to the third equation, we eliminate x_2 from the third equation. This produces the reduced system of equations

$$\begin{bmatrix} 0.000003 & 0.213472 & 0.332147 \\ 0 & -15334.9 & -23860.0 \\ 0 & 0 & -0.500000 \end{bmatrix} \begin{bmatrix} x_1 \\ x_2 \\ x_3 \end{bmatrix} \doteq \begin{bmatrix} 0.235262 \\ -16900.5 \\ -0.200000 \end{bmatrix} \tag{2.10}$$

which has a triangular coefficient matrix.

If we use back substitution (that is, solve the equations in the order 3, 2, 1) we obtain the computed solution

$$\hat{x} = \begin{bmatrix} -1.33333 \\ 0.479723 \\ 0.400000 \end{bmatrix} . \tag{2.11}$$

*In order to keep the notation simple, we do not use superscripts to indicate that a_{22} and a_{32} are from (2.8), not (2.5).

To see how bad this solution is we observe that x (correct to 10 significant digits) is

$$(2.12) \qquad x \doteq \begin{bmatrix} -0.9912894252 \\ 0.05320393391 \\ 0.6741214694 \end{bmatrix} .$$

Wilkinson shows that a simple modification of Gaussian elimination called <u>partial pivoting for size</u> will correct the difficulty demonstrated in this example. The principle involved in this modification is to select the largest element (in magnitude) among a_{11}, a_{21}, and a_{31} as the first pivot and then to interchange the appropriate pair of equations so that the first pivot will always be the element (in the first column of the coefficient matrix) with greatest magnitude. In this example, then, (2.5) should be replaced by

$$(2.13) \qquad \begin{bmatrix} 0.215512 & 0.375623 & 0.476625 \\ 0.000003 & 0.213472 & 0.332147 \\ 0.173257 & 0.663257 & 0.625675 \end{bmatrix} \begin{bmatrix} x_1 \\ x_2 \\ x_3 \end{bmatrix} = \begin{bmatrix} 0.127653 \\ 0.235262 \\ 0.285321 \end{bmatrix}$$

before the first step of Gaussian elimination is attempted.

Notice that (2.13) is mathematically equivalent to (2.5) since an interchange of the first two equations corresponds to an interchange of the first two rows of the augmented matrix [A,b].

From a mathematical point of view the row interchange has accomplished nothing, but from a computational point of view the interchange alters the values of the multipliers m_{21} and m_{23} and this implies a change in the algorithm.

By using the partial pivotal strategy at each stage of elimination we guarantee that the modulus of each multiplier remains less than or equal to unity. That this modified algorithm is more numerically stable (for this example) than unmodified Gaussian elimination is demonstrated by the fact that the computed solution becomes

$$\hat{x} = \begin{bmatrix} -0.991291 \\ 0.0532050 \\ 0.674122 \end{bmatrix} . \tag{2.14}$$

What caused the unmodified Gaussian elimination to yield such poor results (2.11)? The answer is quite clear when we examine the computation in some detail. In (2.5) we observe that a_{11} is extremely small relative to the other elements in the coefficient matrix which implies that m_{21} and m_{31} will be quite large. In fact, we find the computed values to be

$$\hat{m}_{21} = -71837.3 \tag{2.15}$$

and

$$\hat{m}_{31} = -57752.3 . \tag{2.16}$$

In order to compute the values of a_{21}, a_{22}, and a_{23} in the reduced system (2.8), using six-digit floating-point arithmetic, we observe that

$$\begin{aligned} \hat{a}_{21} &= fl[0.215512 - (71837.3)(0.000003)] \\ &= fl[0.215512 - fl(71837.3)(0.000003)] \\ &= fl[0.215512 - 0.215512] \\ &= 0.000000 , \end{aligned} \tag{2.17}$$

and this is the correct answer to six significant digits. Moreover,

(2.18)
$$\begin{aligned}\hat{a}_{22} &= fl[0.375623 - (71837.3)(0.213472)] \\ &= fl[0.375623 - fl(71837.3)(0.213472)] \\ &= fl[0.375623 - 15335.3] \\ &= -15334.9\end{aligned}$$

Notice that the last step required that we subtract two numbers, each containing six significant digits. However, because one number was quite large relative to the other, we could only take advantage of one significant digit of accuracy in the smaller number. Thus, we actually carried out the computation

(2.19)
$$0.4 - 15335.3 = -15334.9$$

and this would have been the same result if the first number on the left had been <u>any</u> number x in the range

(2.20)
$$0.350000 < x < 0.450000.$$

A similar difficulty is encountered in the computation of a_{23} (see Young and Gregory [1973], p. 807).

1 REMARK This loss of accuracy which occurs when a very large number is added to or subtracted from a very small number is the second source of possible computational innacuracy we have encountered. The first occurred in Chapter 1, Section 5, when the subtraction of two nearly equal numbers resulted in cancellation.

Wilkinson [1965] has shown that there are examples for which the partial pivoting strategy in selecting the pivots is not sufficiently stable for some systems of equations and he recommends complete pivoting for size in those cases. The principle here is to select as the first pivot the largest element (in magnitude) in the entire coefficient matrix A . To place this element in the pivotal position we must not only interchange two rows of the augmented matrix [A,b], but we must interchange two columns of A as well.

For example, if a_{23} has the greatest magnitude among the elements of the coefficient matrix in the system

$$\begin{bmatrix} a_{11} & a_{12} & a_{13} \\ a_{21} & a_{22} & a_{23} \\ a_{31} & a_{32} & a_{33} \end{bmatrix} \begin{bmatrix} x_1 \\ x_2 \\ x_3 \end{bmatrix} = \begin{bmatrix} b_1 \\ b_2 \\ b_3 \end{bmatrix} , \tag{2.22}$$

then

$$\begin{bmatrix} a_{23} & a_{22} & a_{21} \\ a_{13} & a_{12} & a_{11} \\ a_{33} & a_{32} & a_{31} \end{bmatrix} \begin{bmatrix} x_3 \\ x_2 \\ x_1 \end{bmatrix} = \begin{bmatrix} b_2 \\ b_1 \\ b_3 \end{bmatrix} \tag{2.23}$$

would be the mathematically equivalent system after complete pivoting. Notice that the components in the solution vector must be permuted in exactly the same manner as the permutation in the columns of A .

CHAPTER 3

A SCALING PROBLEM

Wilkinson has suggested [1961, p. 284] that before any attempt is made to solve a large system of linear algebraic equations, numerically, the system should be scaled in some manner. For example, the equations

$$(1.1)\qquad \begin{bmatrix} 1,000,000 & 990,000 \\ 0.99 & 0.98 \end{bmatrix} \begin{bmatrix} x_1 \\ x_2 \end{bmatrix} = \begin{bmatrix} 1,990,000 \\ 1.97 \end{bmatrix}$$

are mathematically equivalent to the equations

$$(1.2)\qquad \begin{bmatrix} 1.00 & 0.99 \\ 0.99 & 0.98 \end{bmatrix} \begin{bmatrix} x_1 \\ x_2 \end{bmatrix} = \begin{bmatrix} 1.99 \\ 1.97 \end{bmatrix}$$

but the "scaled" system (1.2) is preferred whenever a numerical solution is to be attempted. This preference is not merely for esthetic reasons because Wilkinson [1965, p. 194] points out that it is possible to decrease the condition number* of the problem by scaling (a useful device if the problem is ill-conditioned).

Strangely enough the scaling problem for linear equations is more difficult to solve than it would appear at first glance. Wilkinson [1965] uses the term <u>equilibrated</u> to describe a system of equations $Ax = b$ in which each row (and each column) has been scaled to have a length of order unity. The term is not used too precisely, however, and in practice an equilibrated matrix $A = (a_{ij})$ is one whose rows

---

*The condition number has been defined as $N = \|A\| \cdot \|A^{-1}\|$.

and columns have been scaled so that $|a_{ij}| \leq 1$, for all i and j, and each row (and each column) has at least one element greater than or equal to $1/\beta$, where β is the radix for floating-point computation on the digital computer used.

1.3 REMARK In order to avoid introducing rounding errors in the process of scaling, it is recommended that all scale factors be powers of β. This is extremely important if the problem is ill-conditioned.

One of the first things one observes is that the equilibrated form of a matrix is not unique and, consequently, different equilibrated forms have different condition numbers. The following example due to Hamming appears in Forsythe and Moler [1967, p.45].

$$A = \begin{bmatrix} 1 & 1 & 2(10^9) \\ 2 & -1 & 10^9 \\ 1 & 2 & 0 \end{bmatrix}. \tag{1.4}$$

If the rows are scaled by powers of ten, we obtain the equilibrated matrix

$$A' = \begin{bmatrix} 10^{-10} & 10^{-10} & 0.2 \\ 2(10^{-9}) & -10^{-9} & 1.0 \\ 0.1 & 0.2 & 0.0 \end{bmatrix} \tag{1.5}$$

and we do not need column scaling. On the other hand, if the columns are scaled first, we obtain the equilibrated matrix

$$A'' = \begin{bmatrix} 0.1 & 0.1 & 0.2 \\ 0.2 & -0.1 & 0.1 \\ 0.1 & 0.2 & 0.0 \end{bmatrix} \tag{1.6}$$

and we do not need row scaling. The matrices A' and A'' have different condition numbers and the systems of equations corresponding to these two matrices have a different choice of pivots for Gaussian elimination.

Scaling, therefore, needs to be examined with some care. It is well known that scaling rows and columns of a matrix A can be accomplished by pre- and post- multiplying A by appropriate diagonal matrices. In order to relate this to the system of equations $Ax = b$, let D_1 and D_2 be non-singular diagonal matrices and let

$$b = D_1 b' \tag{1.7}$$

and

$$x = D_2 x' \tag{1.8}$$

If these quantities are substituted into $Ax = b$, we obtain

$$A(D_2 x') = D_1 b'. \tag{1.9}$$

Therefore,

$$(D_1^{-1} A D_2) x' = b', \tag{1.10}$$

which can be written

$$A' x' = b', \tag{1.11}$$

if we define

(1.12) $$A' = D_1^{-1} A D_2 ,$$

is the scaled equivalent of $Ax = b$ because we can solve (1.10) for x' and use (1.7) to obtain x .

The big questions, of course, are how to select D_1 and D_2 and for what objective. An obvious answer to the second question is to choose D_1 and D_2 in order to reduce the condition number of A . Forsythe and Moler [1967, pp.42-43] suggest that the questions be phrased as follows:

(i) Given a matrix A , what choice of nonsingular diagonal matrices D_1 and D_2 will cause the condition number of $A' = D_1^{-1} A D_2$ to be a minimum?

(ii) If minimizing diagonal matrices D_1 and D_2 exist, in answer to the previous question, can they (or reasonable approximations) be computed using a sufficiently fast computer algorithm?

Since the condition number $N = \|A\| \cdot \|A^{-1}\|$ depends on the particular matrix norm used, the answer to (i) depends on our selection of a norm. Bauer [1963] has solved the problem for the maximum norm $\|\cdot\|_\infty$. However, no solution has been found for an arbitrary matrix norm.

In order to present Bauer's solution for the maximum norm we must first introduce some terminology. A matrix A is called <u>partly decomposable</u> if and only if there exist permutation matrices P and Q such that

$$PAQ = \begin{bmatrix} M & R \\ O & N \end{bmatrix}, \tag{1.13}$$

where M and N are square and O is null. If $Q = P^T$, then A is called <u>decomposable</u>. If A is not decomposable it is called <u>indecomposable</u>.

Notice that every decomposable matrix is also partly decomposable. However, some indecomposable matrices are not partly decomposable. Consequently, we need one additional term. If A is not partly decomposable, it is called <u>fully indecomposable</u>. See Figure 1.14 below.

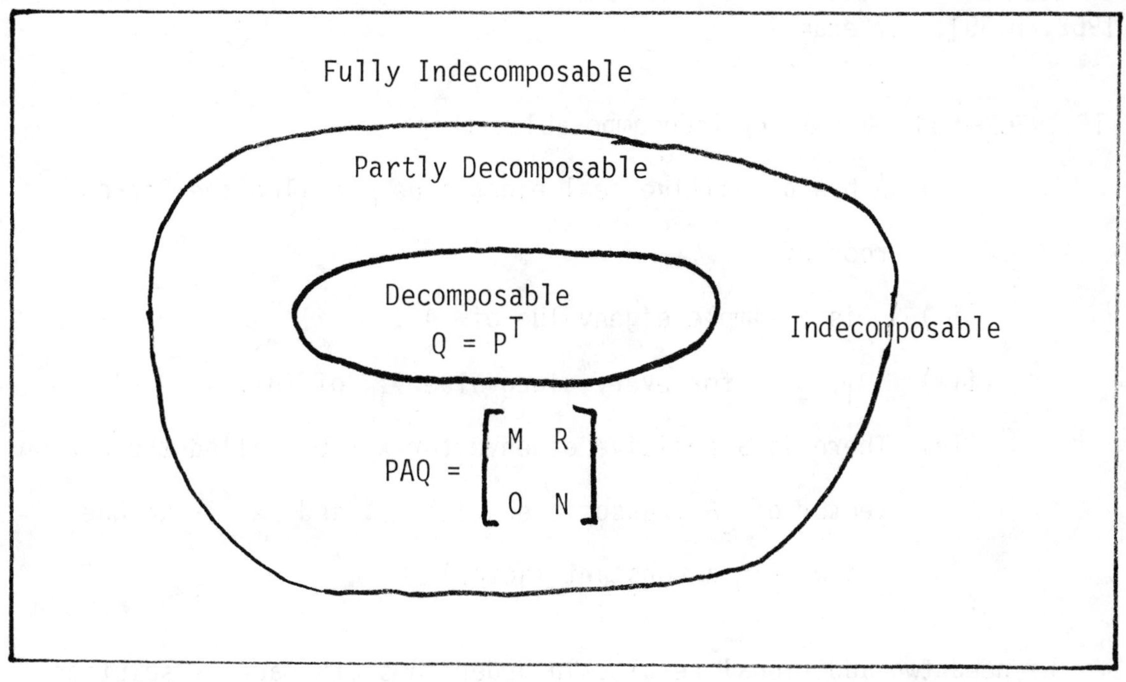

1.14 Figure.

1.15 <u>REMARK</u> The decomposability concept is very useful. For example, if A is decomposable, then $Ax = b$ can be replaced by

$$\begin{bmatrix} M & R \\ O & N \end{bmatrix} \begin{bmatrix} x' \\ x'' \end{bmatrix} = \begin{bmatrix} b' \\ b'' \end{bmatrix}$$

by an appropriate interchange of rows and columns of A corresponding to an interchange* of equations and unknowns. This is equivalent to the two smaller systems

$$\begin{cases} Nx'' = b'' \\ Mx' = b' - Rx'' . \end{cases}$$

If $A \in \mathbf{R}^{nn}$, and if $a_{ij} \geq 0$ for all i and j, we call $A = (a_{ij})$ a <u>nonnegative</u> matrix and write $A \geq 0$. When these are strict inequalities we say that A is a <u>positive</u> matrix. With this background we can state the well known Perron-Frobenius theorem. See Varga [1962, p.30], for example.

1.16 <u>THEOREM</u> If $A \geq 0$ is indecomposable, then

(i) A has a positive real eigenvalue ρ called the Perron root of A.

(ii) ρ is a simple eigenvalue of A.

(iii) $|\lambda_i| \leq \rho$ for every eigenvalue λ_i of A.

(iv) There is a positive eigenvector $x > 0$ called the Perron vector of A, associated with ρ, and x is unique (to within a constant factor).

We need two additional results in order to state Bauer's solution to the scaling problem. See Businger [1968, p.348].

1.17 <u>THEOREM</u> If A is nonsingular and fully indecomposable, then $|A| \cdot |A^{-1}|$ is indecomposable.

---

*Recall (2.22) and (2.23) of Chapter 2.

1.18 <u>THEOREM</u> Under the same hypothesis, $|A^{-1}|\cdot|A|$ is indecomposable.

We can now formulate Bauer's solution to the scaling problem for the maximum norm $\|\cdot\|_\infty$. Let A be nonsingular and fully indecomposable and form the two nonnegative matrices

(1.19) $$U = |A|\cdot|A^{-1}|$$

and

(1.20) $$V = |A^{-1}|\cdot|A| .$$

From the last two theorems above, both U and V are indecomposable. Since U and V have the same set of eigenvalues (See Young and Gregory [1973, p.755], for example) it follows from the Perron-Frobenius theorem that there exists a Perron root ρ common to both U and V. Their Perron vectors are not common, however, and we denote them by u and v, respectively.

Suppose we let

(1.21) $$u = \begin{bmatrix} u_1 \\ u_2 \\ \vdots \\ u_n \end{bmatrix} \quad \text{and} \quad v = \begin{bmatrix} v_1 \\ v_2 \\ \vdots \\ v_n \end{bmatrix}$$

(scale factors are not relevant here) and form the diagonal matrices

(1.22) $$D_1 = \begin{bmatrix} u_1 & & & \\ & u_2 & & \\ & & \ddots & \\ & & & u_n \end{bmatrix}$$

and

$$(1.23) \qquad D_2 = \begin{bmatrix} v_1 & & & \\ & v_2 & & \\ & & \ddots & \\ & & & v_n \end{bmatrix} .$$

Then Bauer shows that, with respect to the maximum norm, the condition number of $D_1^{-1}AD_2$ is a minimum. In fact, he shows that this minimum value is ρ, that is,

$$(1.24) \qquad \rho = \| D_1^{-1}AD_2\|_\infty \cdot \| D_2^{-1}A^{-1}D_1\|_\infty .$$

1.25 REMARK Bauer's solution is of great theoretical interest, of course, but it is not too practical because one has to solve a matrix eigenvalue problem (a major problem in its own right) merely to scale a system of linear equations prior to attempting a solution. This demonstrates our claim that scaling is not a simple procedure.

For example, the matrix

$$(1.26) \qquad A = \begin{bmatrix} a & b \\ c & d \end{bmatrix}$$

is nonsingular and fully indecomposable if $ad \neq bc$ and if $a,b,c,d \in \mathbb{R}^+$. In this case,

$$(1.27) \qquad A^{-1} = \frac{1}{\det A} \begin{bmatrix} d & -b \\ -c & a \end{bmatrix}$$

and so

$$(1.28) \qquad U = \frac{1}{|\det A|}\begin{bmatrix} (ad + bc) & 2ab \\ 2cd & (ad + bc) \end{bmatrix}$$

and

$$(1.29) \qquad V = \frac{1}{|\det A|}\begin{bmatrix} (ad + bc) & 2bd \\ 2ac & (ad + bc) \end{bmatrix}.$$

The common Perron root of U and V is

$$(1.30) \qquad \rho = \frac{ad + bc + 2\sqrt{abcd}}{|ad - bc|}$$

and the Perron vectors for U and V are

$$(1.31) \qquad u = \begin{bmatrix} \sqrt{abcd} \\ cd \end{bmatrix}$$

and

$$(1.32) \qquad v = \begin{bmatrix} \sqrt{abcd} \\ ac \end{bmatrix},$$

respectively. Thus,

$$(1.33) \qquad D_1 = \begin{bmatrix} \sqrt{abcd} & o \\ o & cd \end{bmatrix}$$

and

$$(1.34) \qquad D_2 = \begin{bmatrix} \sqrt{abcd} & o \\ o & ac \end{bmatrix}.$$

Consequently,

$$D_1^{-1}AD_2 = \begin{bmatrix} a & \sqrt{\frac{abc}{d}} \\ \sqrt{\frac{abc}{d}} & a \end{bmatrix} \tag{1.35}$$

is the optimally scaled matrix (with respect to the maximum norm) corresponding to A .

It is interesting to note that if

$$A = \begin{bmatrix} 100 & 0.01 \\ 99 & 0.01 \end{bmatrix}, \tag{1.36}$$

the condition number is

$$\begin{aligned} N &= \| A\|_\infty \cdot \| A^{-1} \|_\infty \\ &= 1{,}990{,}199 \end{aligned} \tag{1.37}$$

whereas $D_1^{-1}AD_2$ has a condition number slightly less than 400. This demonstrates the power of optimum scaling.

1.38 <u>REMARK</u> Because of the complexity of Bauer's algorithm for optimum scaling with respect to the maximum norm $\| \cdot \|_\infty$ and because of the fact that no one has solved the problem for arbitrary norms, there is no known practical solution to the scaling problem in general. However, McKeeman [1962] suggests a fairly simple scaling procedure which is in common use. He suggests scaling the rows of A to the same maximum (for example, divide each equation by the coefficient of greatest absolute value). In practice, though, he suggests using scale factors which are powers of β (the radix) so as to avoid introducing rounding errors during the scaling process.

CHAPTER 4

THE USE OF RESIDUE ARITHMETIC FOR OBTAINING EXACT COMPUTATIONAL RESULTS

1. Introduction

In Chapter 1 we discussed the difficulties associated with attempts at using the finite system of floating-point numbers $(\mathbf{F}, +, \cdot)$ and an automatic digital computer to simulate arithmetic in the field of real numbers $(\mathbf{R}, +, \cdot)$. Since the data encountered in many problems consists of sets of rational numbers, it is often the case that we are trying to simulate arithmetic in the field of rational numbers $(\mathbf{Q}, +, \cdot)$. However, in either case, the result of the simulation is inexact arithmetic and the introduction of rounding errors.

Because of the difficulties associated with inexact arithmetic and because of the difficulties associated with attempts at solving ill-conditioned problems using inexact arithmetic, there is a strong motivation to investigate finite number systems for doing exact arithmetic using an automatic digital computer. The ability to do exact arithmetic will enable us to avoid completely some of the difficulties mentioned in the previous chapters.

It is well known that automatic digital computers can perform certain arithmetic operations exactly if the operands are integers. Hence, it is valid to ask whether or not there exist classes of problems for which the data describing the problems consist of sets of integers and for which most (or even all) computational steps used in obtaining a solution can be carried out using integer arithmetic. Fortunately such problems exist; for example, systems of linear algebraic equations whose coefficients are rational numbers. By properly scaling the rational coefficients we can replace them by integers. For example, the system

$$(1.1)\qquad \begin{bmatrix} \frac{1}{4} & \frac{2}{3} \\ \frac{1}{2} & \frac{1}{3} \end{bmatrix} \begin{bmatrix} x_1 \\ x_2 \end{bmatrix} = \begin{bmatrix} \frac{11}{12} \\ \frac{5}{6} \end{bmatrix}$$

is mathematically equivalent to the system

$$(1.2)\qquad \begin{bmatrix} 3 & 8 \\ 3 & 2 \end{bmatrix} \begin{bmatrix} x_1 \\ x_2 \end{bmatrix} = \begin{bmatrix} 11 \\ 5 \end{bmatrix}.$$

For a treatment of this class of problems by the methods described in this chapter see Young and Gregory [1973], Chapter 13.

2. Single-Modulus Residue Arithmetic

Since we are interested in using a finite number system to do exact arithmetic and since arithmetic using integer operations can be done exactly in certain cases, it is clear that we should examine residue or modular arithmetic. For example, the set of integers $\{0,1,2,\cdots,p-1\}$ where p is a prime, constitutes a finite field, the Galois field $GF(p)$, under addition and multiplication modulo p. Even if we choose a modulus m, not a prime, we obtain a finite commutative ring under addition and multiplication modulo m. If we do integer arithmetic and reduce the results modulo m, we are using an arithmetic system called <u>single-modulus residue arithemtic</u>, where the integer m is called the <u>modulus</u> of the arithmetic system.

Theoretical background

Many of the results which follow are found in Young and Gregory [1973], Chapter 13. Those proofs of theorems which are omitted here can be found in Szabó and Tanaka [1967]. Throughout this book we use the symbol **I** to denote the set of integers.

Let $a, b, m \in \mathbf{I}$, with[†] $m > 1$. Then, if m divides $a-b$, we write

$$a \equiv b \pmod{m} \tag{2.1}$$

and say a is congruent to b modulo m. If m does not divide $a-b$, we write

$$a \not\equiv b \pmod{m} \tag{2.2}$$

and say a is not congruent to b modulo m.

[†]Negative integers can be used as moduli but to no apparent advantage, since m divides $a-b$ if and only if $-m$ divides $a-b$.

2.3 Theorem. Let a, b, c, d, x, y, m $\in \mathbf{I}$, with $m > 1$. Then

(a) The following statements are equivalent

(i) $$a \equiv b \pmod{m}$$

(ii) $$b \equiv a \pmod{m}$$

(iii) $$a - b \equiv 0 \pmod{m} .$$

(b) If

$$a \equiv b \pmod{m}$$

and

$$b \equiv c \pmod{m} ,$$

then

$$a \equiv c \pmod{m} .$$

(c) If

$$a \equiv b \pmod{m}$$

and

$$c \equiv d \pmod{m} ,$$

then

$$ax + cy \equiv bx + dy \pmod{m} .$$

(d) If

$$a \equiv b \pmod{m}$$

and

$$c \equiv d \pmod{m} ,$$

then

$$ac \equiv bd \pmod{m} .$$

In single-modulus residue arithmetic we can represent each $a \in \mathbf{I}$ by an integer r , called its <u>residue</u> modulo m , where r satisfies the relations

(2.4) $$a \equiv r \pmod{m}$$

and

$$(2.5) \qquad 0 \leq r < m \ .$$

A common notational convention (see Szabó and Tanaka [1967], for example) is to represent r by writing

$$(2.6) \qquad r = |a|_m \ .$$

For example, if $a = 58$ and $m = 5$, then $|58|_m = 3$ and we say that 58 has been <u>reduced</u> modulo 5 .

2.7 Theorem. Let $a, b, c, m \in \mathbf{I}$, with $m > 1$. Then

(a) $$|a|_m \text{ is unique.}$$

(b) $$|a|_m = |b|_m \text{ if and only if } a \equiv b \pmod m \ .$$

(c) $$|km|_m = 0 \text{ for every } k \in \mathbf{I} \ .$$

(d)
$$\begin{aligned} |a + b|_m &= \Big| |a|_m + |b|_m \Big|_m \\ &= \Big| |a|_m + b \Big|_m \\ &= \Big| a + |b|_m \Big|_m \ . \end{aligned}$$

(e)
$$\begin{aligned} |ab|_m &= \Big| |a|_m |b|_m \Big|_m \\ &= \Big| |a|_m \, b \Big|_m \\ &= \Big| a|b|_m \Big|_m \ . \end{aligned}$$

Notice that (d) and (e) describe addition and multiplication modulo m . It is clear that there is complete freedom of choice as to when to reduce an operand modulo m . For example, if we use (d) , we see that there are four ways to add 15 and 23 modulo 7 .

(i) $$|15 + 23|_7 = |38|_7 = 3 ,$$

(ii) $$|15 + 23|_7 = |1 + 2|_7 = 3 ,$$

(iii) $$|15 + 23|_7 = |15 + 2|_7 = 3 ,$$

and

(iv) $$|15 + 23|_7 = |1 + 23|_7 = 3 .$$

In order to describe subtraction and division modulo m we first examine the (finite) set of non-negative residues modulo m ,

(2.8) $$\mathbf{I}_m = \{0,1,2,\cdots,m-1\} .$$

2.9 Theorem. The system $(\mathbf{I}_m,+,\cdot)$, where + and $\cdot$ denote addition modulo m and multiplication[†] modulo m , respectively, constitutes a finite commutative ring.

Proof We merely verify that the following properties are valid for every $a,b,c \in \mathbf{I}_m$.

[†]We often omit the symbol $\cdot$ and write ab instead of $a \cdot b$.

closure	$\lvert a+b\rvert_m \in \mathbf{I}_m$	$\lvert ab\rvert_m \in \mathbf{I}_m$
commutativity	$\lvert a+b\rvert_m = \lvert b+a\rvert_m$	$\lvert ab\rvert_m = \lvert ba\rvert_m$
associativity	$\lvert a+(b+c)\rvert_m = \lvert (a+b)+c\rvert_m$	$\lvert a(bc)\rvert_m = \lvert (ab)c\rvert_m$
unique identities	$\lvert a+0\rvert_m = \lvert a\rvert_m$	$\lvert a\cdot 1\rvert_m = \lvert a\rvert_m$
unique inverse	$\lvert a+\underline{a}\rvert_m = 0$	. . . .
distributivity	$\lvert a(b+c)\rvert_m = \lvert ab+ac\rvert_m$	

where $\underline{a}$, the <u>additive inverse of a modulo m</u>, is

$$\underline{a} = \lvert -a\rvert_m .$$

We can define subtraction in the ring $(\mathbf{I}_m,+,\cdot)$ as the addition of the additive inverse. Thus,

$$\text{(2.10)} \qquad \lvert a - b\rvert_m = \lvert a + \underline{b}\rvert_m = \Big\lvert a + \lvert -b\rvert_m \Big\rvert_m .$$

We can define division, when it exists, as multiplication by the multiplicative inverse. However, the big question is: "When does it exist?" The following theorem will help us answer the question.

2.11 <u>Theorem</u>. The finite commutative ring $(\mathbf{I}_m,+,\cdot)$ is a finite field, if and only if, m is a prime.

<u>Proof</u>. See McCoy [1948], page 22, for example.

Consequently, if m is a prime, $(\mathbf{I}_m,+,\cdot)$ is the Galois field GF(m) and every non-zero element in $\mathbf{I}_m$ has a <u>multiplicative inverse modulo m</u> which is defined as follows. If $a \neq 0$ and $a \in \mathbf{I}_m$, there exists a unique integer $b \in \mathbf{I}_m$ which satisfies

$$(2.12) \qquad |ab|_m = |ba|_m = 1.$$

We call b the multiplicative inverse of a modulo m and write

$$(2.13) \qquad b = a^{-1}(m).$$

If m is not a prime, $(\mathbf{I}_m,+,\cdot)$ is not a field and a non-zero element may or may not have a multiplicative inverse. To know when one exists we need the following theorem and corollary. The expression (a,b) denotes the greatest common divisor of a and b.

2.14 <u>Theorem</u>. Let $a \in \mathbf{I}$. Then there exists a unique integer $b \in \mathbf{I}_m$ which satisfies

$$|ab|_m = |ba|_m = 1,$$

if and only if, $|a|_m \neq 0$ and $(a,m) = 1$.

2.15 <u>Corollary</u>. If $a \in \mathbf{I}_m$ is non-zero, then $a^{-1}(m) \in \mathbf{I}_m$ exists (and is unique), if and only if, a and m are relatively prime.

For example, if $m = 10$, $\mathbf{I}_{10} = \{0,1,2,3,4,5,6,7,8,9\}$ and only 1,3,7 and 9 have multiplicative inverses modulo 10. These are 1,7,3 and 9, respectively. On the other hand, if $m = 5$ (a prime), then $\mathbf{I}_5 = \{0,1,2,3,4\}$ and all non-zero elements (that is, 1,2,3 and 4) have multiplicative inverses modulo 5. These are 1,3,2 and 4, respectively. Obviously, $(\mathbf{I}_5,+,\cdot)$ is the Galois field GF(5).

In summary, then, the system $(\mathbf{I}_m,+,\cdot)$ is always a finite commutatative ring and, in particular, if m is a prime, it is also a finite field, the Galois field GF(m). In either case, if $q^{-1}(m)$ exists, we define division† modulo m as follows:

(2.16)
$$\left|\frac{p}{q}\right|_m = |p\cdot q^{-1}|_m$$

We should point out that the quotient of two integers in single-modulus residue arithmetic, when it exists, is always an integer, even in those cases where q does not divide p.

2.17 Example.

$$\left|\frac{4}{7}\right|_{10} = \left|4\cdot 7^{-1}\right|_{10}$$
$$= |4\cdot 3|_{10}$$
$$= 2 \ .$$

In this example, it appears that single-modulus residue arithmetic cannot be used to divide the integer 4 by the integer 7. However, it is not correct to say that the result of this computation is

---

†If the modulus is clearly understood, we write q^{-1} instead of $q^{-1}(m)$.

meaningless, as the following example demonstrates.

2.18 Example.

$$\left|\frac{4}{7}\cdot 7\right|_{10} = \left|\left|\frac{4}{7}\right|_{10}\cdot 7\right|_{10}$$

$$= |2\cdot 7|_{10}$$

$$= 4 .$$

Thus, the integer we computed in Example 2.17 can be used as an intermediate result in the calculation above. This illustrates the fact that single-modulus residue arithmetic can be used in carrying out a sequence of arithmetic operations on integers in $\mathbf{I}_m$ even though the sequence involves one or more division operations, as long as m is relatively prime to each integer which appears in a denominator (so that the appropriate multiplicative inverses modulo m exist).

The only difficulty is in interpreting the computed results. If the correct answer is an integer in $\mathbf{I}_m$, then the result obtained using residue arithmetic will agree with the correct answer. If, on the other hand, the correct answer is not an integer in $\mathbf{I}_m$, then the result obtained using residue arithmetic will not agree with the correct answer and some additional information will be needed in order to obtain the correct answer. See Problem 2.28 below, for example.

2.19 Remark. It is obvious that the modulus m should be a prime, if this is at all feasible, because this guarantees that $(\mathbf{I}_m,+,\cdot)$ is the Galois field GF(m) so that all non-zero elements in $\mathbf{I}_m$ will have multiplicative inverses modulo m.

Applications of the theory

We have defined $|a|_m$ in (2.6) to be the least non-negative residue modulo m. With this definition, computation in single-modulus residue arithmetic is simple, in the sense that only positive integers are involved in our system $(\mathbf{I}_m,+,\cdot)$. However, if we wish to solve problems such as the determinant evaluation problem described in (1.3), where some of the matrix elements are negative integers, we must be able to handle negative integers as well as positive integers.

This can be done by introducing a system of symmetric residues modulo m. For symmetry with respect to the origin, m must be an odd integer. In this case we retain (2.4) but replace (2.5) by the inequality

(2.20) $$-\frac{m}{2} < r < \frac{m}{2} .$$

Thus, along with the complete set of non-negative residues, $\mathbf{I}_m$, we have the complete set of symmetric residues,

(2.20) $$\mathbf{S}_m = \left\{-\frac{m-1}{2}, \cdots, -2, -1, 0, 1, 2, \cdots, \frac{m-1}{2}\right\}.$$

A common notational convention (see Howell and Gregory [1970], page 24) for the symmetric residue of a modulo m is

$$r = /a/_m. \tag{2.22}$$

It is easily verified that $(\mathbf{S}_m,+,\cdot)$ is a finite commutative ring and, in particular, if m is a prime, $(\mathbf{S}_m,+,\cdot)$ is a finite field. It is also easily verified that $(\mathbf{S}_m,+,\cdot)$ is isomorphic to $(\mathbf{I}_m,+,\cdot)$. Consequently, if the data describing a problem consist of integers in $\mathbf{S}_m$, we can map them into $\mathbf{I}_m$, carry out the computations[†], and map the results back into $\mathbf{S}_m$. The mapping functions for doing this are

$$|a|_m = \begin{cases} /a/_m \,, & \text{if } 0 \le /a/_m < \frac{m}{2} \\ /a/_m + m \,, & \text{otherwise,} \end{cases} \tag{2.23}$$

and

$$/a/_m = \begin{cases} |a|_m \,, & \text{if } 0 \le |a|_m < \frac{m}{2} \\ |a|_m - m \,, & \text{otherwise.} \end{cases} \tag{2.24}$$

Figure 2.25 illustrates the relationship between $\mathbf{S}_m$ and $\mathbf{I}_m$ for the case $m = 11$.

[†]All computations could be done in $(\mathbf{S}_m,+,\cdot)$, of course. This requires that algebraic signs be monitored, however.

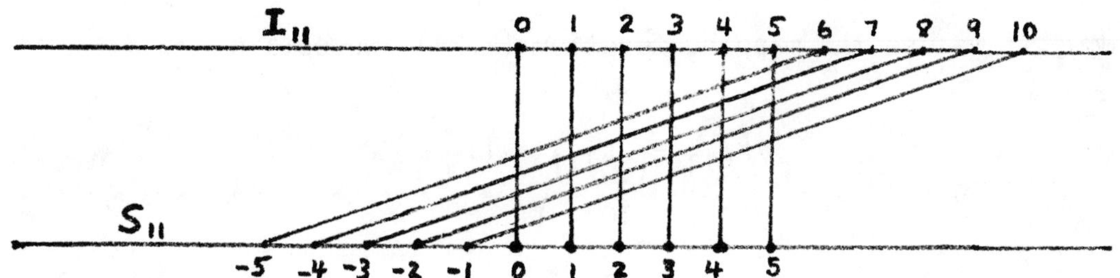

2.25 Figure. The mappings between $\mathbf{S}_{11}$ and $\mathbf{I}_{11}$.

2.26 Problem. Evaluate x , if

$$x = \frac{2}{3} - \frac{5}{3}$$

$$= \frac{2}{3} + \frac{(-5)}{3} \quad .$$

Solution.

We choose $m = 11$ and use the mapping function (2.23) to write

$$|-5|_{11} = /-5/_{11} + 11$$

$$= 6 \ .$$

Thus, in GF(11) ,

$$|x|_{11} = \left|\frac{2}{3} + \frac{6}{3}\right|_{11}$$

$$= \left|2 \cdot 3^{-1} + 6 \cdot 3^{-1}\right|_{11}$$

$$= |(2)(4) + (6)(4)|_{11}$$

$$= \left|8 + |24|_{11}\right|_{11}$$

$$= |8 + 2|_{11}$$

$$= 10 \ .$$

Now, if we use the mapping function (2.24), we obtain

$$\begin{aligned} /x/_{11} &= |x|_{11} - 11 \\ &= 10 - 11 \\ &= -1 \end{aligned}$$

and this is the correct answer.

In Problem 2.26 x is an integer. Consequently, we can choose a modulus m in such a way that single-modulus residue arithmetic gives us the correct answer. However, suppose we have a problem for which is not an integer but the rational number $x = p/q$, where q does not divide p. In this case we have two choices; either we organize the computation in such a way that p and q are computed independently, or we compute the integer $\left|\frac{p}{q}\right|_m = |x|_m$ and recover the correct value of x by using additional information.

For example, if we should happen to know kq, a multiple of q, and $kp, kq \in S_m$, then $x = \frac{kp}{kq}$ where

(2.27) $$kp = /kq|x|_m/_m.$$

The following problem demonstrates this technique.

2.28 <u>Problem</u>. Evaluate x, if

$$\begin{aligned} x &= \frac{1}{2} - \frac{2}{3} - \frac{1}{6} \\ &= \frac{1}{2} + \frac{(-2)}{3} + \frac{(-1)}{6} \end{aligned}$$

<u>Solution</u>

We choose $m = 19$ and use the mapping function (2.23) to write

$$\begin{aligned} |-2|_{19} &= /-2/_{19} + 19 \\ &= 17, \end{aligned}$$

and

$$\left|-1\right|_{19} = /-1/_{19} + 19$$

$$= 18.$$

Thus, in GF(19),

$$|x|_{19} = \left|\frac{1}{2} + \frac{17}{3} + \frac{18}{6}\right|_{19}$$

$$= \left|2^{-1} + |17 \cdot 3^{-1}|_{19} + |18 \cdot 6^{-1}|_{19}\right|_{19}$$

$$= \left|10 + |(17)(13)|_{19} + |(18)(16)|_{19}\right|_{19}$$

$$= |10 + 12 + 3|_{19}$$

$$= 6$$

Now the least common denominator of the three fractions is 6, so x can be written as a fraction whose denominator is 6. Thus, we know that $kq = 6$. We, also, know that $|x|_{19} = 6$. With these two values in (2.27), we obtain

$$kp = /kq|x|_{19}/_{19}$$

$$= /(6)(6)/_{19}$$

$$= -2$$

Since $x = \frac{kp}{kq}$ we have

$$x = \frac{-2}{6}$$

$$= -\frac{1}{3},$$

and this is the correct answer.

2.29 Remark. Every integer which does not lie in $\mathbf{S}_m$ is congruent modulo m to a unique integer which does lie in $\mathbf{S}_m$. Consequently, if the result of a computation is an integer which is not an element of $\mathbf{S}_m$, then single-modulus residue arithmetic will not produce the correct integer but an integer congruent to the correct integer. This is called overflow. Obviously, if we wish to prevent overflow, we must be sure that, for any given calculation, the modulus m is large enough to guarantee that $\mathbf{S}_m$ contains the result.

3. Using the Euclidean Algorithm to Compute $a^{-1}(m)$.

We have used the multiplicative inverse, $a^{-1}(m)$, in our examples above but no algorithm for computing $a^{-1}(m)$ has been presented. We shall present one in this section. We recall, from Corollary 2.15, that if $a \in \mathbf{I}_m$ is non-zero, then $a^{-1}(m) \in \mathbf{I}_m$ exists (and is unique) if and only if, a and m are relatively prime.

Let $(a,m) = d$. Then integers w and z exist which enable us to write

$$aw + mz = d \tag{3.1}$$

or, more appropriately,

$$aw = d - mz. \tag{3.2}$$

Then, from Theorem 2.7,

$$\begin{aligned} |aw|_m &= |d - mz|_m \\ &= |d|_m. \end{aligned} \tag{3.3}$$

Thus, whenever a and m are relatively prime, that is, whenever d = 1,

$$|w|_m = a^{-1}(m). \tag{3.4}$$

The Euclidean algorithm enables us to compute d and, through the process, derive what we need to compute w and, subsequently, $a^{-1}(m)$. To do this we write

$$\text{(3.5)} \quad \left\{ \begin{array}{ll} m = aq_1 + r_1 & 0 < r_1 < a \\ a = r_1q_2 + r_2 & 0 < r_2 < r_1 \\ r_1 = r_2q_3 + r_3 & 0 < r_3 < r_2 \\ \vdots & \vdots \\ r_{n-3} = r_{n-2}q_{n-1} + r_{n-1} & 0 < r_{n-1} < r_{n-2} \\ r_{n-2} = r_{n-1}q_n + r_n & 0 < r_n < r_{n-1} \\ r_{n-1} = r_nq_{n+1} & \end{array} \right.$$

where the greatest common divisor is the final non-zero remainder. Thus, $d = r_n$. We can obtain w and z by eliminating $r_1, r_2, \ldots, r_n$ from (3.5).

3.6 Example a = 35, m = 47. If we use (3.5), we obtain

$$47 = (35)(1) + 12$$

$$35 = (12)(2) + 11$$

$$12 = (11)(1) + 1$$

$$11 = (1)(11).$$

Since the final non-zero remainder $r_3 = 1 = d$, 35 and 47 are relatively prime. Hence, $|w|_m = a^{-1}(m)$ in this example. We can find w and z by eliminating r_1, r_2, and r_3. Thus from (3.5),

$$\begin{aligned} d &= r_3 \\ &= r_1 - r_2 q_3 \\ &= r_1 - (a - r_1 q_2) q_3 \\ &= r_1 (1 + q_2 q_3) + a(-q_3) \\ &= m(1 + q_2 q_3) + a(-q_1 - q_3 - q_1 q_2 q_3) \end{aligned}$$

and, using (3.1), we have w and z by inspection. Therefore,

$$\begin{aligned} a^{-1}(m) &= |w|_m \\ &= |-q_1 - q_3 - q_1 q_2 q_3|_m . \end{aligned}$$

In our numerical example $q_1 = 1$, $q_2 = 2$, and $q_3 = 1$ so that

$$w = -4 .$$

Thus,

$$\begin{aligned} 35^{-1}(47) &= |-4|_{47} \\ &= 43. \end{aligned}$$

We can carry out this computation in a mechanical fashion, easy to program for a digital computer, by following the steps indicated in the logical flow chart which follows.[†]

[†] We use $\lfloor x \rfloor$ to denote the largest integer smaller than or equal to x.

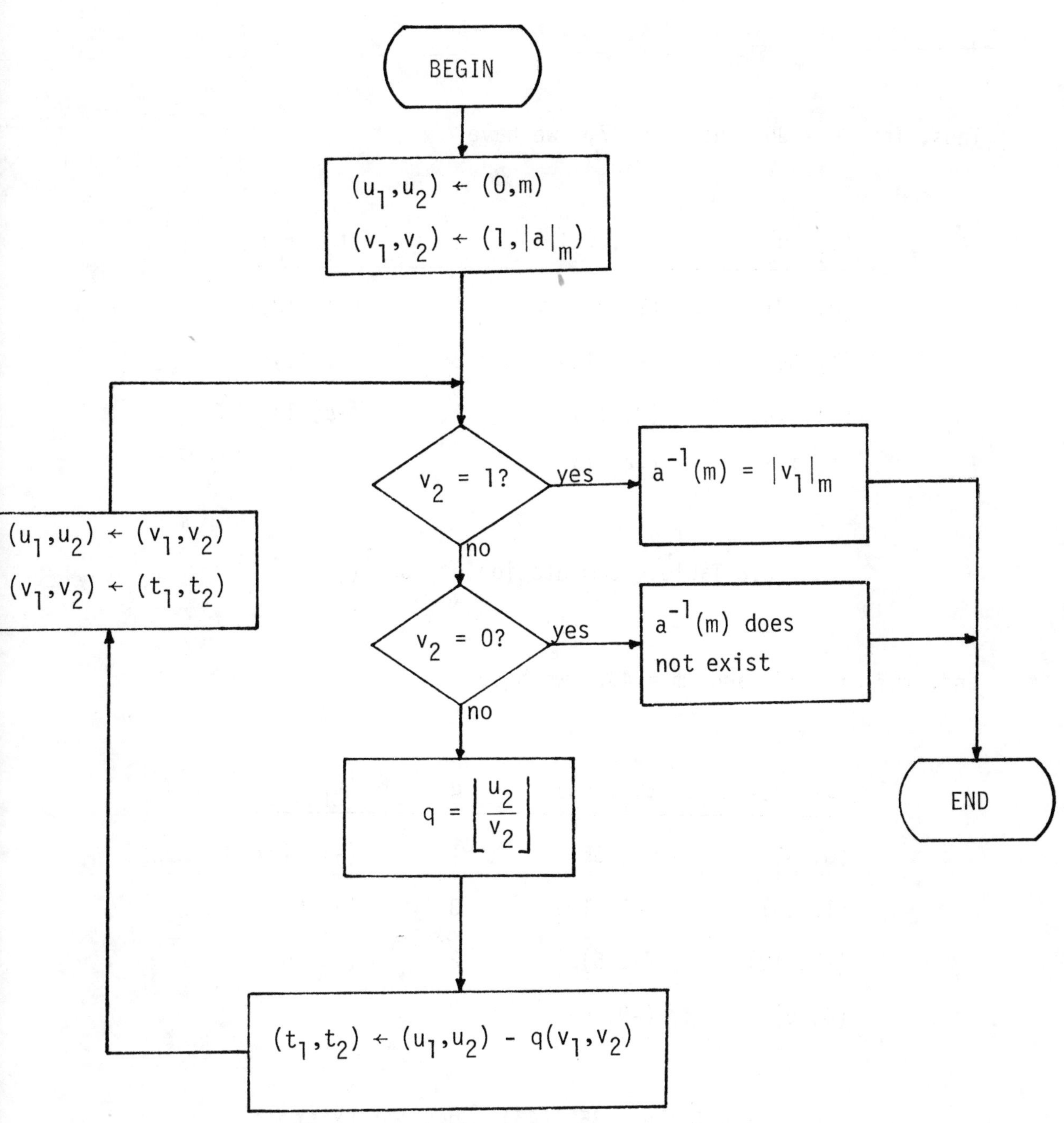

3.7 Figure†. An Algorithm for $a^{-1}(m)$.

---

†This flowchart was suggested by S.C. Ong.

Thus, for $a = 35$ and $m = 47$, we have

(u_1, u_2)	(v_1, v_2)	q	(t_1, t_2)
(0, 47)	(1, 35)	1	(-1, 12)
(1, 35)	(-1, 12)	2	(3, 11)
(-1, 12)	(3, 11)	1	(-4, 1)
(3, 11)	(-4, 1)		

3.8 Table. Computation of $35^{-1}(47)$

and, for $a = 35$ and $m = 45$, we have

(u_1, u_2)	(v_1, v_2)	q	(t_1, t_2)
(0, 45)	(1, 35)	1	(-1, 10)
(1, 35)	(-1, 10)	3	(4, 5)
(-1, 10)	(4, 5)	2	(-9, 0)
(4, 5)	(-9, 0)		

3.9 Table. $35^{-1}(45)$ Does Not Exist

Notice that $(a, m) = d$ is generated in both Table 3.8 and Table 3.9. In the first example $(35, 47) = 1$ and this shows up as the last value of v_2 in Table 3.8. In the second example $(35, 45) = 5$ and this shows up as the last non-zero value of v_2 in Table 3.9.

When $(a, m) = 1$, that is, when a and m are relatively prime, as in Table 3.8, we know that $a^{-1}(m) = |w|_m$ because of (3.4). It is not difficult to show that $v_2 = 1$ implies $v_1 = w$ so that in Table 3.8

(3.10) $$\begin{aligned}(v_1, v_2) &= (-4, 1)\\ &= (w, 1),\end{aligned}$$

and $w = -4$. Therefore

(3.11) $$\begin{aligned}35^{-1}(47) &= |-4|_{47}\\ &= 43,\end{aligned}$$

the same result we obtained above.

4. Multiple-Modulus Residue Arithmetic

The best way to handle the problem of overflow, described in Remark 2.2 is to use more than one modulus. This is due to the fact that multiple-modulus residue arithmetic using several moduli can be shown to be equivalent to single-modulus residue arithmetic using the least common multiple of the moduli as the single modulus.

For example, consider the ordered n-tuple

$$\beta = [m_1, m_2, \ldots, m_n], \tag{4.1}$$

whose components are the (distinct) moduli $m_1, m_2, \ldots, m_n$. Assume that the moduli are pairwise relatively prime, that is, that

$$(m_i, m_j) = 1 \qquad i \neq j. \tag{4.2}$$

Whenever β in (4.1) satisfies (4.2) we call it a <u>base vector</u> for the residue number systems we are about to describe.

4.3 <u>Definition.</u> For each integer s we call the (unique) ordered n-tuple of residues

$$|s|_\beta = \left[|s|_{m_1}, |s|_{m_2}, \ldots, |s|_{m_n}\right]$$

the <u>standard[†] residue representation</u> of s with respect to the base vector β. The individual residues $|s|_{m_i}$ are called the <u>standard residue digits</u> of s with respect to β.

4.4 <u>Example.</u> Let $\beta = [5,7,9]$ and $s = 34$. Then

$$|34|_\beta = [4,6,7].$$

[†]We use the word "standard" to distinguish these representations from the "symmetric" residue representations mentioned in Remark 4.14.

Let M be the product of the moduli in the base vector β, that is, let

$$M = \prod_{i=1}^{n} m_i . \qquad (4.5)$$

We are now in a position to state an important theorem and its corollary. In both the theorem and the corollary M should be replaced by the least common multiple of the moduli. However, since the moduli are assumed to be pairwise relatively prime, M is the least common multiple.

4.6 Theorem. Two integers s and t have the same standard residue representation with respect to β, that is, $|t|_\beta = |s|_\beta$, if and only if,

$$s \equiv t \pmod{M}.$$

4.7 Corollary. If $t = |s|_M$, then t and s have the same standard residue representation with respect to β, that is, $|t|_\beta = |s|_\beta$.

4.8 Example. Let β = [3,5,7] so that M = 105. Also, let s = 403. Then

$$|403|_{105} = 88 .$$

It is easily verified (by direct computation) that

$$|403|_\beta = |88|_\beta$$
$$= [1,3,4].$$

Suppose we consider, for each $a \in \mathbf{I}$, the unique integer r which satisfies

(4.9) $$a \equiv r \pmod{M}$$

and

(4.10) $$0 \leq r < M \quad .$$

Then

(4.11) $$r = |a|_M$$

belongs to the set of non-negative residues modulo **M**

(4.12) $$\mathbf{I}_M = \{0,1,2, \ldots, M-1\}$$

4.13 <u>Definition</u>. <u>The standard residue number system</u> for the base vector β is the M — member set of standard residue representations

$$\mathbf{I}_\beta = \left\{ |s|_\beta \ : \ s \in \mathbf{I} \right\} .$$

Obviously, $\mathbf{I}_\beta$ is a finite number system and its elements are in one-to-one correspondence with the elements of $\mathbf{I}_M$ from Theorem 4.6 and Corollary 4.7. As a consequence of Theorem 2.9, the system $(\mathbf{I}_M, +, \cdot)$, where + and $\cdot$ denote addition and multiplication modulo M , constitutes a finite commutative ring. It is not difficult to show that we can define binary operations $\boxplus$ and $\boxdot$ on elements of $\mathbf{I}_\beta$ in such a way that the system ($\mathbf{I}_\beta$, $\boxplus$, $\boxdot$) also constitutes a finite commutative ring. (See Theorem 4.15 and Theorem 4.16, for example).

It can be shown that the rings ($\mathbf{I}_M$, +, $\cdot$) and ($\mathbf{I}_\beta$, $\boxplus$, $\boxdot$) are isomorphic and, as a consequence, multiple-modulus residue arithmetic is equivalent to single-modulus residue arithmetic with M as the modulus.

4.14 Remark. If $\beta = [m_1, m_2, \ldots, m_n]$ contains only odd moduli, we can use (2.20) and (2.22) to define the corresponding symmetric residue number system for the base vector β. In this case we represent an integer s by the symmetric residue representation.

$$/s/_\beta = \left[/s/_{m_1}, /s/_{m_2}, \ldots, /s/_{m_n}\right],$$

where the individual residues $/s/_{m_i}$ are the symmetric residue digits. We define the symmetric residue number system to be the M - member set of symmetric residue representations

$$\mathbf{S}_\beta = \left\{/s/_\beta : s \in \mathbf{I}\right\}.$$

In $\mathbf{S}_\beta$ every integer s is represented by a unique n-tuple $/s/_\beta$ and the correspondence is one-to-one for those integers in the set

$$\mathbf{S}_M = \left\{-\frac{M-1}{2}, \ldots, -2, -1, 0, 1, 2, \ldots, \frac{M-1}{2}\right\}.$$

Analogous to $(\mathbf{I}_M, +, \cdot)$ and $(\mathbf{I}_\beta, \boxplus, \boxdot)$ we have the finite commutative rings $(\mathbf{S}_M, +, \cdot)$ and $(\mathbf{S}_\beta\ \boxplus, \boxdot)$. It can be shown that they are isomorphic and, indeed, that the latter two rings are isomorphic with the former two. Thus, $\mathbf{I}_\beta$, $\mathbf{S}_\beta$, $\mathbf{I}_M$, and $\mathbf{S}_M$ are all M-member sets and we have a one-to-one correspondence between the standard and the symmetric residue number systems.

Note that for any integer a, conversions between $|a|_\beta$ and $/a/_\beta$ are accomplished by the mappings (2.23) and (2.24) applied to corresponding residue digits $|a|_{m_i}$ and $/a/_{m_i}$ for $i = 1,2, \ldots, n$.

Arithmetic in $\mathbf{I}_\beta$

We describe addition, subtraction, and multiplication in $\mathbf{I}_\beta$ as follows:

4.15 Theorem. Let a and b be integers. The standard residue representation of the integer $a \pm b$ with respect to the base vector β is

$$|a \pm b|_\beta = \left[|z_1|_{m_1}, |z_2|_{m_2}, \ldots, |z_n|_{m_n} \right]$$

where

$$z_i = |a|_{m_i} \pm |b|_{m_i} \qquad i = 1,2, \ldots, n.$$

4.16 <u>Theorem</u>. Let a and b be integers. The standard residue representation of the integer ab with respect to the base vector β is

$$|ab|_\beta = \left[|w_1|_{m_1}, |w_2|_{m_2}, \ldots, |w_n|_{m_n} \right]$$

where

$$w_i = |a|_{m_i} |b|_{m_i} \qquad i = 1,2, \ldots, n.$$

Thus, to get the residue digits of a sum, a difference, or a product of two integers we merely add, subtract, or multiply, respectively, the corresponding residue digits of the two operands in componentwise fashion and reduce each result modulo the appropriate modulus.

4.17 <u>Example</u>. Let $\beta = [3, 5, 7]$ with $M = 105$. Also, let $a = 24$ and $b = 20$. Then

$$|24|_\beta = [0, 4, 3]$$

and

$$|20|_\beta = [2, 0, 6].$$

Thus,

$$|24 + 20|_\beta = \left[|0+2|_3, |4+0|_5, |3+6|_7 \right]$$

$$= [2, 4, 2].$$

Similarly,

$$|24 - 20|_\beta = \left[|0-2|_3, |4-0|_5, |3-6|_7 \right]$$

$$= [1, 4, 4],$$

and

$$|(24)(20)|_\beta = \left[|(0)(2)|_3 , |(4)(0)|_5 , |(3)(6)|_7 \right]$$
$$= [0, 0, 4].$$

As a check we observe that

$$|44|_\beta = [2, 4, 2]$$
$$|4|_\beta = [1, 4, 4]$$
$$|480|_\beta = [0, 0, 4]$$

4.18 Remark. In the three computations just described notice that the correct sum, difference, and product of 24 and 20 are 44, 4, and 480, respectively. The first two results are elements of the set $\mathbf{I}_{105}$ but the third is not. Thus, the representations [2, 4, 2] and [1, 4, 4] will be correctly mapped[†] onto 44 and 4, respectively, but [0, 0, 4] will be mapped onto 60 (not 480) since

$$0 \le 60 < 105$$

and

$$480 \equiv 60 \pmod{105}.$$

Notice that M = 105 was not large enough to prevent "overflow" in the case of the product. The only thing to do, of course, is to make M larger either by choosing *more* moduli or by choosing *larger* moduli.

[†] The method used to map residue representations onto integers is described in the next section.

In order to do division in the standard residue number system we have a slight generalization of the procedure used in single-modulus residue arithmetic.

4.19 Definition. Let a be an integer and let $a^{-1}(m_1)$, $a^{-1}(m_2)$, ..., $a^{-1}(m_n)$ exist. Then the n-tuple

$$a^{-1}(\beta) = \left[a^{-1}(m_1),\ a^{-1}(m_2),\ \ldots,\ a^{-1}(m_n)\right]$$

is called the standard residue representation of the multiplicative inverse of a with respect to the base vector β.

For the existence of $a^{-1}(\beta)$ we merely require that $a^{-1}(m_i)$ exist for $i = 1,2, \ldots, n$. Obviously, when $a^{-1}(\beta)$ exists, it is unique.

4.20 Theorem. Let a and b be integers and let $a^{-1}(\beta)$ exist. Then

$$\left|\frac{b}{a}\right|_{\beta} = \left[|b\cdot a^{-1}(m_1)|_{m_1},\ |b\cdot a^{-1}(m_2)|_{m_2},\ \ldots,\ |b\cdot a^{-1}(m_n)|_{m_n}\right].$$

Notice that

$$(4.21) \qquad \left|\frac{a}{a}\right|_{\beta} = [1,\ 1,\ \ldots,\ 1]$$

as it should.

4.22 Example. Let $\beta = [3,5,7]$ with $M = 105$, and let $a = 23$ and $b = 46$. In this example a divides b. Since

$$|46|_\beta = [1,1,4]$$

and

$$23^{-1}(\beta) = [2,2,4],$$

we have

$$\left|\frac{46}{23}\right|_\beta = \left[|1\cdot 2|_3 , |1\cdot 2|_5 , |4\cdot 4|_7\right]$$
$$= [2,2,2]$$

and this is the correct representation for 2.

4.23 Remark. If a does not divide b we have a result analogous to the situation associated with single-modulus residue arithmetic. The answer may be difficult to interpret (without further information) but it is valid as an intermediate result for further computation.

5. Mapping Standard Residue Representations Onto Integers

In Remark 4.18 we indicated that, for $\beta = [3,5,7]$, the residue representations [2,4,2], [1,4,4], and [0,0,4] were mapped onto the integers 44, 4, and 60, respectively, but we gave no indication as to how the mappings were carried out. The purpose of this section is to answer the question "how do we map a standard residue representation (with respect to β) onto a unique integer in $\mathbf{I}_M$?"

One of the oldest known algorithms (but not the fastest) for carrying out this mapping makes use of a classical theorem from the theory of numbers called the Chinese Remainder Theorem. (See Young and Gregory [1973, p. 874], for example.) We shall not describe that algorithm here, however. Instead, we present an algorithm which uses a mixed-radix number representation for each integer.

Consider the ordered n-tuple

$$\rho = [r_1, r_2, \ldots, r_n] \tag{5.1}$$

where the components $r_1, r_2, \ldots, r_n$ are called radices. Let R be the product of the radices, that is,

$$R = \prod_{i=1}^{n} r_i \; . \tag{5.2}$$

It is well known (see Szabó and Tanaka [1967, p. 41], for example) that every integer s in the range

$$0 \leq s < R \tag{5.3}$$

can be expressed uniquely in the form

(5.4) $$s = d_0 + d_1(r_1) + d_2(r_1r_2) + \dots + d_{n-1}(r_1r_2\dots r_{n-1}),$$

where d_0, d_1, ..., d_{n-1} are the <u>standard mixed-radix digits</u> satisfying the inequalities

(5.5) $$0 \le d_i < r_{i+1} \qquad i = 0,1,\dots, n-1.$$

Notice that the primary role played by r_n is to establish a bound on d_{n-1}.

The <u>digit sequence</u> for s in this mixed-radix representation is the ordered set of digits $d_0, d_1, \dots, d_{n-1}$ which we display in the form

(5.6) $$\langle s \rangle_\rho = \langle d_0, d_1, \dots, d_{n-1} \rangle$$

For example, if $\rho = [2,3,5]$, then $R = 30$. Consequently, since

(5.7) $$29 = 1 + 2(2) + 4(2\cdot 3),$$

we know that

(5.8) $$\langle 29 \rangle_\rho = \langle 1, 2, 4 \rangle .$$

5.9 <u>Definition</u>. The <u>standard mixed-radix system</u> for ρ is the set of digit sequences $\langle s \rangle_\rho$ for integers in the range

$$0 \le s < R .$$

A special case occurs if $r_1 = r_2 = \dots = r_n$, the familiar <u>fixed-radix</u> number representation. If each radix is ten, for example, this is merely the decimal representation. A more interesting special case (for our purpos occurs if $r_i = m_i$ for $i = 1, 2, \ldots, n$, where m_i is an element of the base vector β for our multiple-modulus residue system, and r_i is an element of ρ defined in (5.1). In this case

(5.10) $$\rho = \beta ,$$

and so the ranges for the multiple-modulus residue system and its associated mixed-radix system are both $0 \leq s < M$. This is extremely important since we shall wish to change from one system to the other.

Consider the standard mixed-radix system and its associated standard residue system for the base vector $\beta = [m_1, m_2, \ldots, m_n]$. If an integer s has the representation

(5.12) $$s = d_0 + d_1(m_1) + d_2(m_1m_2) + \ldots + d_{n-1}(m_1m_2 \ldots m_{n-1}),$$

with the (unique) digit sequence

(5.13) $$\langle s \rangle_\beta = \langle d_0, d_1, \ldots, d_{n-1} \rangle$$

in the former, and the residue representation

(5.14) $$|s|_\beta = \left[|s|_{m_1}, |s|_{m_2}, \ldots, |s|_{m_n} \right]$$

in the latter, we see from (5.5) and (2.5) that both the mixed-radix digits d_{i-1} and the residue digits $|s|_{m_i}$ lie in the same closed interval $[0, m_i-1]$ for $i = 1, 2, \ldots, n$.

Suppose we are given the residue representation $|s|_\beta$ in (5.14) and we wish to find $\langle s \rangle_\beta$ in (5.13). In other words, suppose we are given the residue digits $|s|_{m_i}$ and we wish to find the mixed-radix digits d_{i-1} for $i = 1, 2, \ldots, n$. To get d_0 let $s = t_1$ and observe that, from (5.12),

(5.15) $$t_1 = s$$

$$= d_0 + m_1[d_1 + d_2(m_2) + \ldots + d_{n-1}(m_2 \ldots m_{n-1})]$$

$$= d_0 + m_1 t_2 .$$

Hence, from Theorem 2.7,

(5.16) $$|t_1|_{m_1} = |d_0 + m_1 t_2|_{m_1}$$

$$= d_0 .$$

Notice that $d_0 = |t_1|_{m_1} = |s|_{m_1}$, which implies that the first mixed-radix digit equals the first residue digit and <u>no computation is necessary</u>.

To compute d_1 we use (5.15) and write

(5.17) $$t_2 = d_1 + m_2[d_2 + d_3(m_3) + \ldots + d_{n-1}(m_3 \ldots m_{n-1})]$$

$$= d_1 + m_2 t_3 .$$

Hence, from Theorem 2.7,

(5.18) $$|t_2|_{m_2} = |d_1 + m_2 t_3|_{m_2}$$

$$= d_1 .$$

This suggests the following recursion for obtaining the mixed-radix digits. Beginning with the initial values $t_1 \equiv s$ and $d_0 = |t_1|_{m_1}$ we successively compute t_{i+1} and d_i (in that order) using the equations

$$(5.19)\qquad \begin{cases} t_{i+1} = \dfrac{t_i - d_{i-1}}{m_i} \\ d_i = |t_{i+1}|_{m_{i+1}} \end{cases} \qquad i = 1, 2, \ldots, n-1.$$

The computation indicated in (5.19) for computing the mixed-radix digits can be carried out using residue arithmetic. This is a critical point because <u>it cannot be carried out using ordinary arithmetic</u>. The reason for this is that we must know s to get started (recall our definition, $t_1 = s$) and we do not know s. (That is what we are trying to find.) However, we do know $|s|_\beta$ and that is all we need when we use residue arithmetic.

<u>The computation of d_i using residue arithmetic</u>

Let $|s|_\beta$ be defined by (5.14). If we set $s = t_1$ and use (5.16) we can write

$$(5.20)\qquad |t_1|_\beta = \left[d_0\ ,\ |t_1|_{m_2}\ ,\ \ldots\ ,\ |t_1|_{m_n}\right].$$

By definition,

$$(5.21)\qquad |d_0|_\beta = \left[|d_0|_{m_1}\ ,\ |d_0|_{m_2}\ ,\ \ldots\ ,\ |d_0|_{m_n}\right],$$

which allows us to compute

$$(5.22)\qquad |t_1 - d_0|_\beta = \left[0,\ |z_2|_{m_2}\ ,\ |z_3|_{m_3},\ \ldots\ ,\ |z_n|_{m_n}\right],$$

with

$$z_i = |t_1|_{m_i} - |d_0|_{m_i} \qquad i = 2,3,\ldots,n. \tag{5.23}$$

If we introduce the reduced base vector

$$\beta_1 = [m_2 , m_3 , \ldots , m_n], \tag{5.24}$$

we can represent $t_1 - d_0$ with respect to β_1 . In this case,

$$|t_1 - d_0|_{\beta_1} = \left[|z_2|_{m_2}, |z_3|_{m_3} , \ldots , |z_n|_{m_n}\right]. \tag{5.25}$$

In order to compute t_2 we need $m_1^{-1}(\beta_1)$. This multiplicative inverse exists because m_1 is relatively prime to each of the elements of β_1. Thus,

$$m_1^{-1}(\beta_1) = \left[m_1^{-1}(m_2) , m_1^{-1}(m_3) , \ldots , m_1^{-1}(m_n)\right] \tag{5.26}$$

and so

$$\begin{aligned} |t_2|_{\beta_1} &= |(t_1 - d_0)/m_1|_{\beta_1} \\ &= \left[|w_2|_{m_2} , |w_3|_{m_3} , \ldots, |w_n|_{m_n}\right], \end{aligned} \tag{5.27}$$

with

(5.28) $$w_i = |z_i|_{m_i}\, m_1^{-1}(m_i) \qquad i = 2,3,\ldots,n.$$

From (5.18) and (5.27) we have

(5.29) $$\begin{aligned}|w_2|_{m_2} &= |t_2|_{m_2} \\ &= d_1 ,\end{aligned}$$

the second mixed-radix digit. If we use this result in (5.27) and note that $|w_i|_{m_i} = |t_2|_{m_i}$, we can write

(5.30) $$|t_2|_{\beta_1} = \left[d_1 , |t_2|_{m_3} , \ldots , |t_2|_{m_n}\right].$$

Analogous to (5.21) we have

(5.31) $$|d_1|_{\beta_1} = \left[|d_1|_{m_2} , |d_1|_{m_3} , \ldots , |d_1|_{m_n}\right]$$

which gives us

(5.32) $$|t_2 - d_1|_{\beta_1} = \left[0 , |v_3|_{m_3} , \ldots , |v_n|_{m_n}\right],$$

with

(5.33) $$v_i = |t_2|_{m_i} - |d_1|_{m_i} \qquad i = 3,4,\ldots,n.$$

If, analogous to β_1, we introduce the reduced base vector

(5.34) $$\beta_2 = [m_3 , m_4 , \ldots , m_n],$$

we can represent $t_2 - d_1$ with respect to β_2. In this case,

(5.35) $$|t_2 - d_1|_{\beta_2} = \left[|v_3|_{m_3}, |v_4|_{m_4}, \ldots, |v_n|_{m_n}\right].$$

In order to compute t_3 we need $m_2^{-1}(\beta_2)$. This multiplicative inverse exists because m_2 is relatively prime to each of the elements of β_2. Thus,

(5.36) $$m_2^{-1}(\beta_2) = \left[m_2^{-1}(m_3), m_2^{-1}(m_4), \ldots, m_2^{-1}(m_n)\right]$$

and so

(5.37) $$\begin{aligned} |t_3|_{\beta_2} &= |(t_2 - d_1)/m_2|_{\beta_2} \\ &= \left[|u_3|_{m_3}, |u_4|_{m_4}, \ldots, |u_n|_{m_n}\right], \end{aligned}$$

with

(5.38) $$u_i = |v_i|_{m_i}\, m_2^{-1}(m_i) \qquad i = 3,4, .$$

From (5.19) we know that

(5.39) $$\begin{aligned} |u_3|_{m_3} &= |t_3|_{m_3} \\ &= d_2, \end{aligned}$$

the third mixed-radix digit. If we continue this algorithm, we eventually compute each of the mixed-radix digits in sequence.

5.40 <u>Problem</u>. If $\beta = [13, 11, 7]$ and $|s|_\beta = [4,2,4]$, what is $\langle s \rangle_\beta$ and what is s ?

<u>Solution</u>

We assume, since $M_\beta = 1001$, that we seek the unique integer in the range $0 \leq s < 1001$. The computation described in (5.20) through (5.39) can be presented in the form of a table.

β	$m_1 = 13$	$m_2 = 11$	$m_3 = 7$	
$\|t_1\|_\beta$	[4]	2	4	
$\|d_0\|_\beta$	4	4	4	subtract
$\|t_1-d_0\|_\beta$	0	9	0	
$m_1^{-1}(\beta_1)$		6	6	multiply
$\|t_2\|_{\beta_1} = \|(t_1-d_0)/m_1\|_{\beta_1}$		[10]	0	
$\|d_1\|_{\beta_1}$		10	3	subtract
$\|t_2-d_1\|_{\beta_1}$		0	4	
$m_2^{-1}(\beta_2)$			2	multiply
$\|t_3\|_{\beta_2} = \|(t_2-d_1)/m_2\|_{\beta_2}$			[1]	

The elements in the dotted squares are d_0, d_1, and d_2, in that order.

Hence, $\langle s \rangle_\beta = \langle 4, 10, 1 \rangle$. Consequently, from (5.12), we obtain

$$\begin{aligned} s &= 4 + 10(13) + 1(13)(11) \\ &= 277, \end{aligned}$$

and this is the correct answer.

5.41 <u>Remark.</u> Since we know how to find the digit sequence $\langle s \rangle_\beta$ from $|s|_\beta$ we merely evaluate the right hand side of (5.12) to find the integer s. In the example above this was quite simple to do directly. However, we use the following recursive procedure to evaluate (5.12), in general.

$$\begin{aligned} S_1 &= d_{n-1} \\ S_2 &= d_{n-2} + S_1 m_{n-1} \\ S_3 &= d_{n-3} + S_2 m_{n-2} \\ &\vdots \\ S_n &= d_0 + S_{n-1} m_1 \end{aligned}$$

and $s = S_n$.

5.42 <u>Remark.</u> At the beginning of Section 4 it was pointed out that multiple modulus residue arithmetic using the base vector β is equivalent to single modulus residue arithmetic using M as the single modulus.† In other

---

† Actually it is the least common multiple of the moduli which is the single modulus. However, since we are assuming that the moduli are pairwise relatively prime, M is the least common multiple.

words, arithmetic in ($\mathbf{I}_\beta$, $\boxplus$, $\boxdot$) is equivalent to arithmetic in (I_M , + , ·).

Suppose we want to do arithmetic on both positive and negative integers. We use the same procedure suggested in Section 2 where integers in $\mathbf{S}_m$ were mapped into $\mathbf{I}_m$ using (2.23) so that arithmetic could be performed in ($\mathbf{I}_m$, + , ·). Results were then mapped back into $\mathbf{S}_m$ using (2.24).

We can do the same thing here. Integers in $\mathbf{S}_M$ can be mapped into $\mathbf{I}_M$ using a mapping similar to (2.23). This allows the arithmetic to be done in ($\mathbf{I}_\beta$, $\boxplus$, $\boxdot$) producing results in $\mathbf{I}_M$. These results can then be mapped into $\mathbf{S}_M$ using a mapping similar to (2.24).

5.43 Remark. We have discussed the four basic arithmetic operations in multiple-modulus residue arithmetic and conversions between residue representations and integers. Notice that arithmetic is done in a componentwise fashion. This suggests that we could take advantage of a computer especially designed for multiple-modulus residue arithmetic. If we had n processors, each designed to carry out residue operations on integers, the operations on the components could be done simultaneously. The newest generation of computers with their vector operations and parallel processors are a step in the right direction. However, none of them has the hardware capability for reducing a result modulo m. It must be done with software.

It should be pointed out, however, that the "First International Workshop on Residue Arithmetic" was held at The Ohio State University on May 31 - June 1, 1978, under the sponsorship of the Department of Electrical Engineering.

At the workshop (most of whose participants were electrical engineers) it was made clear that computers of the next generation would probably have the hardware capability for performing residue arithmetic as one of the options. Predictions were made that computation speeds of the order of 10^9 multiplications per second could be expected using the new technology.

5.44 Remark. We have shown that addition, subtraction and multiplication are extremely simple operations in multiple-modulus residue arithmetic but that division is a bit more complicated. There are three other situations which present us with a bit of complexity:

(i) magnitude comparison for two integers s and t,

(ii) sign detection of s, and

(iii) the recovery of s from $|s|_\beta$.

For a discussion of (i) and (ii) the reader is referred to Szabó and Tanaka [1967], Chapter 4. We have dealt with (iii) in this section and we have discussed (ii) in Remark 5.42. However, our treatment for finding the sign of s involved finding s, itself.

CHAPTER 5

THE USE OF FINITE-SEGMENT P-ADIC ARITHMETIC FOR OBTAINING EXACT COMPUTATIONAL RESULTS

1. Introduction

In the previous chapter we pointed out that one of the advantages of multiple-modulus residue arithmetic is the simplicity of the addition, subtraction, and multiplication algorithms. However, as we indicated, the division algorithm is not very simple because of the difficulties associated with the existence of (and the computation of) the multiplicative inverse. In Remark 5.44 we indicated other difficulties such as sign and magnitude detection (when only $|s|_\beta$ is known) and the problem of recovering s from $|s|_\beta$.

Because of these difficulties we are motivated to present an alternative number system which possesses the best features of both the fixed-radix (or p-ary) number system and the multiple-modulus residue number system. This alternative number system is the finite-segment p-adic number system introduced by Krishnamurthy, Rao, and Subramanian [1975a] and by Alparslan [1975]. This new number system is finite and its relation to the (infinite) p-adic number system of Hensel [1908] is explained in subsequent sections.

In this finite system, each rational number in a certain (finite) set is assigned a unique coded representation called its Hensel code† and arithmetic operations (addition, subtraction, multiplication and division) on pairs of rational numbers in this set can be replaced by corresponding arithmetic operations on their Hensel codes. In particular, the division algorithm is deterministic (free from trial and error) and it is this fact that gives

---

†Named for the German mathematician K. Hensel (1861-1941).

finite-segment p-adic arithmetic its greatest appeal. It should be pointed out that finite-segment p-adic arithmetic shares with residue arithmetic the fact that it is exact.

2. The Field of p-adic Numbers

Let $\mathbf{K}$ be an arbitrary field. We define a norm (sometimes called a valuation) on $\mathbf{K}$ by the following mapping.

2.1 Definition A norm on a field $\mathbf{K}$ (considered a vector space over itself) is a mapping $\|\cdot\|: \mathbf{K} \to \mathbf{R}$ such that, for all α, β in $\mathbf{K}$,

(i) $\|\alpha\| > 0$ and $\|\alpha\| = 0$ if and only if α

(ii) $\|\alpha\beta\| = \|\alpha\| \cdot \|\beta\|$,

(iii) $\|\alpha + \beta\| \leq \|\alpha\| + \|\beta\|$.

For example, in the field of rational numbers $\mathbf{Q}$ the absolute value mapping $|\cdot|$ can be shown to be a norm on $\mathbf{Q}$. Another norm on $\mathbf{Q}$ (of more interest to us here) can be constructed on the observation that, if $\alpha = \frac{a}{b}$ is a non-zero element of $\mathbf{Q}$ with $(a,b) = 1$, then α can be expressed uniquely in the form

$$\alpha = \frac{c}{d} p^e, \tag{2.2}$$

where p is a given prime, c,d, and e are integers with $(c,d) = 1$, and p divides neither c nor d. With α defined in (2.2) we have the following result.

2.3 Theorem. The mapping $\|\cdot\|_p : \mathbf{Q} \to \mathbf{R}$ defined by

$$\|\alpha\|_p = \begin{cases} p^{-e} & \text{if } \alpha \neq 0 \\ 0 & \text{if } \alpha = 0 \end{cases}$$

is a norm on $\mathbf{Q}$.

Proof See Koblitz [1977], page 2.

2.4 Definition. The norm in Theorem 2.3 is called the p-adic norm on $\mathbf{Q}$.

A metric space

Before continuing we need to introduce the concept of a metric and the concept of a metric space.

2.5 Definition. A metric space is a pair $(\mathbf{X}, d)$ consisting of a nonempty set $\mathbf{X}$ and a metric (or distance function) $d: \mathbf{X} \times \mathbf{X} \to \mathbf{R}$ such that, for all x, y, z in $\mathbf{X}$,

(i) $d(x,y) = 0$ if and only if $x = y$

(ii) $d(x,y) = d(y,x)$

(iii) $d(x,z) \leq d(x,y) + d(y,z)$.

The properties (i), (ii), and (iii) are sometimes called the Hausdorff postulates† and it is not difficult to deduce from them a fourth property that, for all x, y in $\mathbf{X}$,

$$\text{(iv)} \quad d(x,y) \geq 0.$$

2.6 <u>Definition</u>. A sequence $(x_n) = (x_1, x_2, \ldots)$, where $x_n \in \mathbf{X}$ for all n, is called a Cauchy sequence‡ in the metric space $(\mathbf{X}, d)$ if and only if

$$d(x_n, x_m) \to 0 \qquad (m, n, \to \infty),$$

that is, for every $\varepsilon > 0$ there exists $N = N(\varepsilon)$ such that for all n, m >

$$d(x_n, x_m) < \varepsilon.$$

2.7 <u>Definition</u>. A sequence $(x_n) = (x_1, x_2, \ldots)$ in the metric space $(\mathbf{X}, d)$ called convergent (to x) if and only if there exists $x \in \mathbf{X}$ such that

$$d(x_n, x) \to 0 \qquad (n \to \infty).$$

We then write $x_n \to x$ and call x the <u>limit</u> of the sequence.

Notice that there is nothing in the definition of a Cauchy sequence in a metric space which implies convergence*. In fact, it is well known that not every Cauchy sequence in a metric space is convergent. However, if a metric space has the property that every Cauchy sequence in the space

---

†Named for the German mathematician F. Hausdorff (1868-1942)

‡Named for the French mathematician A.L. Cauchy (1789-1857)

*Convergence means convergence to a point in the space.

converges, the metric space is special.

2.8 Definition. A metric space ($\mathbf{X}$, d) is called complete if and only if every Cauchy sequence converges (to a point in $\mathbf{X}$). To be more specific, we require that if

$$d(x_n, x_m) \to 0 \qquad (n, m \to \infty),$$

then there exists $x \in \mathbf{X}$ such that

$$d(x_n, x) \to 0 \qquad (n \to \infty).$$

A special metric space

We can construct a metric space by letting $\mathbf{X} = \mathbf{Q}$ and by defining a metric $d: \mathbf{Q} \times \mathbf{Q} \to \mathbf{R}$ in terms of the p-adic norm on $\mathbf{Q}$. Thus, if we define

$$d(\alpha,\beta) = \|\alpha - \beta\|_p \tag{2.9}$$

for all α,β in $\mathbf{Q}$, then ($\mathbf{Q}$, d) constitutes a metric space.

2.10 Definition. The metric (2.9), induced by the p-adic norm $\|\cdot\|_p$, is called the p-adic metric.

Of particular interest in the metric space ($\mathbf{Q}$, d) is the sequence of powers of the prime p

(2.11) $$(p^n) = (p, p^2, p^3, \ldots).$$

It is interesting to note that this sequence converges to 0 because, in terms of the p-adic metric,

(2.12) $$d(p^n, 0) = \|p^n\|_p = p^{-n},$$

and $p^{-n} \to 0$ as $n \to \infty$.

Completion of a metric space

In the theory of metric spaces it is well known that for a non-complete metric space (not every Cauchy sequence converges) it is possible to construct a complete metric space (every Cauchy sequence converges) called the completion of the metric space. See Koblitz [1977], for example.

2.13 Example. Consider the metric space $(\mathbf{Q}, \hat{d})$, where $\hat{d}$ is the absolute value metric

$$\hat{d}(\alpha,\beta) = |\alpha - \beta|.$$

Let $\hat{\mathbf{S}}$ be the set of Cauchy sequences in this metric space. We define two Cauchy sequences $s_1 = (a_1, a_2, \ldots)$ and $s_2 = (b_1, b_2, \ldots)$ to be equivalent (and we write $s_1 \sim s_2$) if and only if $|a_i - b_i| \to 0$ as $i \to \infty$. This is an equivalence relation, that is, $\sim$ has the following properties:

(i) $s \sim s$

(ii) $s_1 \sim s_2$ implies $s_2 \sim s_1$

(iii) $s_1 \sim s_2$ and $s_2 \sim s_3$ imply $s_1 \sim s_3$.

We say that two sequences s_1 and s_2 belong to the same equivalence class if $s_1 \sim s_2$. We now define $\mathbf{R}$ to be the set of equivalence classes of Cauchy sequences in $(\mathbf{Q}, \hat{d})$. It is possible to define addition, multiplication and finding additive and multiplicative inverses of equivalence classes of Cauchy sequences in such a way that $(\mathbf{R}, +, \cdot)$ is a field. (See Koblitz [1977], for example.) This field is the field of real numbers and the metric space $(\mathbf{R}, \hat{d})$ is the completion of the metric space $(\mathbf{Q}, \hat{d})$.

We obtain the field of p-adic numbers if we find the completion of the rational numbers with respect to the p-adic metric rather than the absolute value metric used above. In this case we begin with the metric space $(\mathbf{Q}, d)$, where d is defined in (2.9). We let $\mathbf{Q}_p$ denote the set of equivalence classes of Cauchy sequences in $(\mathbf{Q}, d)$ where s_1 and s_2 are equivalent if and only if $\|a_i - b_i\|_p \to 0$ as $i \to \infty$. With addition, multiplication and finding inverses properly defined (see Koblitz [1977], page 10), the system $(\mathbf{Q}_p, +, \cdot)$ constitutes a field, the field of p-adic numbers, and the metric space $(\mathbf{Q}_p, d)$ is the completion of the metric space $(\mathbf{Q}, d)$.

2.14 Definition. Each element of the set $\mathbf{Q}_p$ is called a p-adic number.

We have introduced the set of p-adic numbers $\mathbf{Q}_p$ in a rather abstract way. Perhaps the following expansion theorem for p-adic numbers will characterize them more concretely. This expansion is somewhat analogous to the decimal expansion for the real numbers.

2.15 Theorem. Any p-adic number $\alpha \in \mathbf{Q}_p$ can be written in the form

$$\alpha = \sum_{j=n}^{\infty} a_j p^j ,$$

where each $a_j \in \mathbf{I}$ and n is such that $\|\alpha\|_p = p^{-n}$. Moreover, if we c each a_j in the interval $[0, p-1]$, then the expansion is unique. (In th case, the expansion is the canonical representation of α).

Proof See Bachman [1964], pages 34-35.

Since the field of rational numbers is imbedded in the field of p-adic numbers, and since we are interested primarily in the p-adic expansion of a rational number, the following corollary is of interest.

2.16 Corollary. Any rational number $\alpha \in \mathbf{Q}$ has the unique p-adic expansio

$$\alpha = \sum_{j=n}^{\infty} a_j p^j$$

where each coefficient a_j is an integer in the interval $[0, p-1]$ and n such that $\|\alpha\|_p = p^{-n}$. (The infinite series converges to the rational n α in the p-adic metric.)

Proof The corollary is a direct consequence of Theorem 2.15.

2.17 Example. Consider the following power series expansion.

$$\begin{aligned}\alpha &= 2 + 3p + p^2 + 3p^3 + p^4 + 3p^5 + \ldots \\ &= 2 + 3p(1 + p^2 + p^4 + \ldots) + p^2(1 + p^2 + p^4 + \ldots) \\ &= 2 + (3p + p^2)(1 + p^2 + p^4 + \ldots).\end{aligned}$$

Since $1 + p^2 + p^4 + \ldots$ converges to $(1 - p^2)^{-1}$ in the p-adic metric we can write, for $p = 5$,

$$\begin{aligned}\alpha &= 2 + \frac{3p + p^2}{1 - p^2} \\ &= 2 - \frac{40}{24} \\ &= \frac{1}{3}\,.\end{aligned}$$

This is usually written in the abbreviated form

$$\frac{1}{3} = .23131313\ldots \qquad (p = 5)$$

where the "point" at the left is called the p-adic point.

2.18 Remark. Notice that there is a one-to-one correspondence between the power series expansion

$$\alpha = a_n p^n + a_{n+1} p^{n+1} + a_{n+2} p^{n+2} + \ldots$$

and the abbreviated representation

$$\alpha = a_n a_{n+1} a_{n+2} \ldots$$

where only the coefficients of the powers of p are exhibited. Because of this correspondence we can use the power series expansion and the abbreviated representation interchangeably. In fact, we shall refer to each of them as the p-adic expansion for α.

The abbreviated representation is completely analogous to the representation of the decimal expansion of a real number. In fact, we complete the analogy by introducing a p-adic point as a device for displaying the sign of n. Thus, we write.

$$\alpha = \begin{cases} a_n a_{n+1} \cdots a_{-2} a_{-1} . a_0 a_1 a_2 \cdots & \text{for } n < 0 \\ . a_0 a_1 a_2 \cdots & \text{for } n = 0 \\ .0 \ldots 0 a_n a_{n+1} \cdots & \text{for } n > 0. \end{cases}$$

We know that a real number is rational if and only if its decimal expansion is periodic. Similarly, a p-adic number is rational if and only if its p-adic expansion is periodic. (See Bachman [1964], page 40.) Consequently, since we are primarily interested in the p-adic expansions of rational numbers we will be dealing mostly with p-adic expansions which are periodic.

The complement representation for a negative rational number

The relationship between the p-adic expansions for α and for $-\alpha$ can be seen in the following result.

2.19 Theorem. If

then
$$\alpha = a_n p^n + a_{n+1} p^{n+1} + a_{n+2} p^{n+2} + \ldots,$$
$$-\alpha = b_n p^n + b_{n+1} p^{n+1} + b_{n+2} p^{n+2} + \ldots,$$

where $b_n = p - a_n$ and $b_j = (p-1) - a_j$ for $j > n$.

Proof We sketch a proof by showing that the sum of the two representations is zero. Write

$$-\alpha = (p-a_n)p^n + (p-1-a_{n+1})p^{n+1} + (p-1-a_{n+2})p^{n+2} + \dots$$

If we form $\alpha + (-\alpha)$ we obtain

$$\begin{aligned} 0 &= p\cdot p^n + (p-1)\cdot p^{n+1} + (p-1)\cdot p^{n+2} + \dots \\ &= 0 + p\cdot p^{n+1} + (p-1)\cdot p^{n+2} + \dots \\ &= 0 + 0 + p\cdot p^{n+2} + \dots \\ &= 0 + 0 + 0 + \dots \end{aligned}$$

with zeros as far as we wish to carry the expansion.

2.20 Example. Recall the 5-adic expansion for 1/3 in Example 2.17.

$$\frac{1}{3} = 2 + 3\cdot 5 + 1\cdot 5^2 + 3\cdot 5^3 + 1\cdot 5^4 + \dots$$

Hence,

$$-\frac{1}{3} = 3 + 1\cdot 5 + 3\cdot 5^2 + 1\cdot 5^3 + 3\cdot 5^4 + \dots$$

and, when we add, we obtain

$$\begin{aligned} 0 &= 5 + 4\cdot 5 + 4\cdot 5^2 + 4\cdot 5^3 + \dots \\ &= 0 + 5\cdot 5 + 4\cdot 5^2 + 4\cdot 5^3 + \dots \\ &= 0 + 0 + 5\cdot 5^2 + 4\cdot 5^3 + \dots \\ &= 0 + 0 + 0 + 5\cdot 5^3 + \dots \\ &= 0 + 0 + 0 + 0 + \dots \end{aligned}$$

The p-adic representation of a radix fraction

If $\alpha = a/b$ is a rational number with $(a,b) = 1$ and if b is a power of p, we call α a radix fraction. It is interesting to observe what happens when the radix fraction is positive.

2.21 Theorem. If $\alpha \in \mathbf{Q}_p$, then α can be represented by a finite p-adic expansion if and only if α is a positive radix fraction.

Proof See MacDuffee [1938].

2.22 Corollary. If α is a positive integer, then α can be represented by a finite p-adic expansion.

Proof If α is a positive integer it is automatically a positive radix fraction.

Notice that when we complement a finite p-adic expansion, using Theorem 2.19, we automatically get an infinite p-adic expansion. Hence, the theorem and corollary above refer only to positive radix fractions (and integers). For example,

(2.23) $$\frac{119}{125} = 442.100000\ldots \qquad (p = 5),$$

and

(2.24) $$-\frac{199}{125} = 102.344444\ldots \qquad (p = 5).$$

It is easily verified that

(2.25) $$119 = .442100000\ldots \qquad (p = 5)$$

and

(2.26) $$-119 = .102344444\ldots \qquad (p = 5)$$

which illustrates the fact that if $\alpha = a/b$, with $(a,b) = 1$ and $b = p^k$, then the p-adic representation of α can be obtained from the p-adic representation of a merely by shifting the p-adic point k places to the right. This is completely analogous to the situation that exists with decimal fractions whose denominators are powers of ten.

2.27 Remark. Since a positive integer h can be expressed in exactly one way as the sum of powers of a prime p, that is, since

$$h = d_0 + d_1p^2 + \ldots + d_kp^k$$

with the integers d_i in the interval $[0, p-1]$, we see that there is essentially no difference between the radix-p (or p-ary) representation of h and the p-adic representation of h. In fact, the only difference is that, in the abbreviated representations, the digits are written in reverse order. For example

$$14 = 2 + 3 + 3^2$$

which means that the radix-3 representation is

$$14_{ten} = 112_{three}\,.$$

However, the 3-adic representation is

$$14_{ten} = .2110000\ldots \qquad (p = 3)$$

and (since the number of non-zero digits is finite) we usually write

$$14_{ten} = .211 \qquad (p = 3).$$

Notice that the radix-3 representation and the 3-adic representation are mirror images of each other.

Computing the digits in a p-adic expansion

Let α have the p-adic expansion

$$(2.28)\qquad \begin{aligned}\alpha &= a_n p^n + a_{n+1}p^{n+1} + a_{n+2}p^{n+2} + \dots \\ &= p^n(a_n + a_{n+1}p + a_{n+2}p^2 + \dots) \\ &= p^n\left[\frac{c_1}{d_1}\right]\end{aligned}$$

where $(c_1,d_1) = 1$ and p divides neither c_1 nor d_1. The p-adic expansion for c_1/d_1 is

$$(2.29)\qquad \frac{c_1}{d_1} = a_n + a_{n+1}p + a_{n+2}p^2 + \dots$$

and so

$$(2.30)\qquad \begin{aligned}\left|\frac{c_1}{d_1}\right|_p &= \left|a_n + a_{n+1}p + a_{n+2}p^2 + \dots\right|_p \\ &= a_n .\end{aligned}$$

In other words, we obtain a_n by computing

$$(2.31)\qquad a_n = \left|c_1 d_1^{-1}\right|_p .$$

Next, we use (2.29) to form the expression

$$\frac{c_1}{d_1} - a_n = p(a_{n+1} + a_{n+2}p + a_{n+3}p^2 + \ldots) = p\left[\frac{c_2}{d_2}\right] \tag{2.32}$$

where $(c_2,d_2) = 1$ and p divides neither c_2 nor d_2. The p-adic expansion for c_2/d_2 is

$$\frac{c_2}{d_2} = a_{n+1} + a_{n+2}p + a_{n+3}p^2 + \ldots \tag{2.33}$$

and so

$$\left|\frac{c_2}{d_2}\right|_p = a_{n+1} . \tag{2.34}$$

In other words, we obtain a_{n+1} by computing

$$a_{n+1} = \left|c_2 d_2^{-1}\right|_p . \tag{2.35}$$

In general, we continue this procedure and, unless α is a positive radix fraction (see Theorem 2.21), the process does not terminate. However, since we are assuming that α is a rational number, the p-adic expansion will be periodic and we need continue only until the period has been exhibited.

2.36 Example. Let $\alpha = 2/3$ and let $p = 5$. In this case $c_1 = 2$, $d_1 = 3$ and $n = 0$. Thus,

$$a_0 = |2 \cdot 3^{-1}|_5$$
$$= 4 .$$

Next we form the expression

$$\frac{c_1}{d_1} - a_0 = \frac{2}{3} - 4$$
$$= 5\left[\frac{-2}{3}\right].$$

In this case $c_2 = -2$ and $d_2 = 3$. Thus,

$$a_1 = |-2 \cdot 3^{-1}|_5$$
$$= 1 .$$

Now we form the expression

$$\frac{c_2}{d_2} - a_1 = \frac{-2}{3} - 1$$
$$= 5\left[\frac{-1}{3}\right].$$

In this case $c_3 = -1$ and $d_3 = 3$. Thus,

$$a_2 = |-1 \cdot 3^{-1}|_5$$
$$= 3.$$

If we contine this procedure we obtain $a_3 = 1$ and $a_4 = 3$. At this point we may stop because the period of the p-adic expansion has been revealed. Hence,

$$\frac{2}{3} = .4131313... \qquad (p = 5).$$

2.37 Remark. We have pointed out some of the similarities between p-adic numbers and decimal numbers. However, one of the differences between p-adic numbers and decimal numbers is that if two p-adic expansions converge to the same number in $\mathbf{Q}_p$, they are the same number, that is, all of their digits are the same. Thus, we never encounter a situation analogous to the example $1 = 0.9999...$, in which a terminating decimal can also be represented by a non-terminating decimal with repeating nines.

3. Arithmetic In $\mathbf{Q}_p$

The operations of addition, subtraction, multiplication and division in $\mathbf{Q}_p$ are quite similar to the corresponding operations on decimals. The main difference, however, is that we proceed from "left to right" rather than from "right to left" as we do with decimals.

Addition

In Example 2.20 we demonstrated how to add 1/3 to -1/3 using their power series expansions. In general, suppose we have two arbitrary p-adic numbers

$$\alpha = a_n p^n + a_{n+1} p^{n+1} + a_{n+2} p^{n+2} + \dots \tag{3.1}$$

and

$$\beta = b_n p^n + b_{n+1} p^{n+1} + b_{n+2} p^{n+2} + \dots \tag{3.2}$$

where the p-adic digits a_i and b_i lie in the interval [0, p-1]. We point out that either a_n or b_n may be zero but not both. There are no conditions imposed on subsequent digits.

If we add α and β we obtain

(3.3)
$$\begin{aligned}\alpha + \beta &= (a_n + b_n)p^n + (a_{n+1} + b_{n+1})p^{n+1} + (a_{n+2} + b_{n+2})p^{n+2} + \dots \\ &= c_n p^n + c_{n+1}p^{n+1} + c_{n+2}p^{n+2} + \dots\end{aligned}$$

where

(3.4)
$$c_i = a_i + b_i, \qquad i = n, n+1, \dots$$

Suppose $c_n, c_{n+1}, \dots, c_{k-1}$ are digits (that is, suppose they are less than p) but c_k is not. Then

(3.5)
$$c_k = p + d_k$$

where $0 \le d_k < p$. In this case,

(3.6)
$$\begin{aligned}\alpha + \beta &= c_n p^n + \dots + c_{k-1}p^{k-1} + (p+d_k)p^k + c_{k+1}p^{k+1} + \dots \\ &= c_n p^n + \dots + c_{k-1}p^{k-1} + d_k p^k + (c_{k+1} + 1)p^{k+1} + \dots\end{aligned}$$

and d_k is the digit associated with p^k. Notice that a "carry" has been generated and so c_{k+1} is increased by one unit. At this point $c_{k+1} + 1$ must be examined to see whether or not it is less than p. If it is not, we have a "carry" propagated to c_{k+2} and so on.

This situation is similar to the situation which exists when we add decimals. However, in the p-adic case we add the digits (using radix-p arithmetic) and, from the point of view of the abbreviated representation, we proceed from left to right (rather than from right to left as in the case of decimals).

3.7 Example. Add 2/3 and 5/6 in $\mathbf{Q}_5$. It is easy to verify that the p-adic expansions for the two operands are

$$\frac{2}{3} = .4131313\ldots \qquad (p = 5)$$

and

$$\frac{5}{6} = .0140404\ldots \qquad (p = 5).$$

Now if we add the p-adic expansions, using radix-5 arithmetic (proceeding from left to right), we obtain

$$\begin{array}{l} .4131313\ldots \\ \underline{.0140404\ldots} \\ .4222222\ldots \end{array}$$

As a check we observe that $2/3 + 5/6 = 3/2$ and

$$\frac{3}{2} = .4222222\ldots \qquad (p = 5).$$

Subtraction

In order to do subtraction we complement the subtrahend (using Theorem 2.19) and add it to the minuend, that is, $\alpha - \beta = \alpha + (-\beta)$.

3.8 Example. Subtract 5/6 from 2/3 in $\mathbf{Q}_5$. If we complement the representation for 5/6 in the previous example, we see that

$$-\frac{5}{6} = .0404040\ldots \qquad (p = 5).$$

Thus, if we add the p-adic expansions for 2/3 and -5/6, using radix-5 arithmetic (proceeding from left to right), we obtain

$$\begin{array}{r} .4131313\ldots \\ \underline{.0404040\ldots} \\ .4040404\ldots \end{array}$$

As a check we observe that $2/3 - 5/6 = -1/6$ and

$$-\frac{1}{6} = .4040404\ldots \qquad (p = 5).$$

Multiplication

Before discussing multiplication we note that we can always write a p-adic number γ in the form

$$\begin{aligned} (3.9) \qquad \gamma &= g_n p^n + g_{n+1} p^{n+1} + g_{n+2} p^{n+2} + \ldots \\ &= p^n (g_n + g_{n+1} p + g_{n+2} p^2 + \ldots) \\ &= p^n \alpha \end{aligned}$$

where α has the obvious definition, and n can be positive, negative, or zero. If we change the notation so that the subscripts on the p-adic digits match the exponents on p, we can write

$$(3.10) \qquad \alpha = a_0 + a_1 p + a_2 p^2 + \ldots$$

3.11 Definition. Any p-adic number whose p-adic expansion contains no negative powers of p is called a p-adic integer. Any p-adic integer whose first digit is non-zero is called a p-adic unit.

Thus, in (3.10) α is a p-adic unit and any p-adic number can be written as a product of a p-adic unit and a power of p.

If we have γ given by (3.9) and

$$\delta = p^m \beta \tag{3.12}$$

where α and β are p-adic units, then

$$\gamma\delta = p^{n+m}\alpha\beta . \tag{3.13}$$

Thus, there is no loss of generality in restricting our discussion of p-adic multiplication to a discussion of the multiplication of p-adic units.

Let α and β be the p-adic units

$$\begin{cases} \alpha = a_0 + a_1 p + a_2 p^2 + \dots \\ \beta = b_0 + b_1 p + b_2 p^3 + \dots \end{cases} \tag{3.14}$$

(where $a_0 b_0 \neq 0$). Then

$$\begin{aligned} \alpha\beta &= (a_0 + a_1 p + a_2 p^2 + \dots)(b_0 + b_1 p + b_2 p^2 + \dots) \\ &= c_0 + c_1 p + c_2 p^2 + c_3 p^3 + \dots \end{aligned} \tag{3.15}$$

with

$$\begin{cases} c_0 = a_0 b_0 \\ c_1 = a_0 b_1 + a_1 b_0 \\ c_2 = a_0 b_2 + a_1 b_1 + a_2 b_0 \\ \vdots \\ c_k = a_0 b_k + a_1 b_{k-1} + \dots + a_k b_0 \\ \vdots \end{cases} \tag{3.16}$$

Even though the p-adic digits a_i and b_i lie in the interval $[0, p-1]$ we cannot assume that the integers c_i lie in this interval, that is, we cannot assume that they are digits. (In general, they are not.) Hence, we write

$$(3.17) \qquad \begin{aligned} c_0 &= a_0 b_0 \\ &= d_0 + t_1 p \end{aligned}$$

where $0 \leq d_0 < p$. Then d_0 is the first digit in the p-adic expansion for $\alpha\beta$ and t_1 is the "carry" which we must add to c_1. Next we write

$$(3.18) \qquad \begin{aligned} c_1 + t_1 &= (a_0 b_1 + a_1 b_0) + t_1 \\ &= d_1 + t_2 p \end{aligned}$$

where $0 \leq d_1 < p$. Then d_1 is the second digit in the p-adic expansion for $\alpha\beta$ and t_2 is the "carry" which we must add to c_2. If we continue this procedure, we obtain the (unique) p-adic expansion

$$(3.19) \qquad \alpha\beta = d_0 + d_1 p + d_2 p^2 + \ldots$$

where $0 \leq d_i < p$ for all i.

Again we point out that this situation is similar to the situation which exists when we multiply decimals. However, in the p-adic case the arithmetic indicated in (3.16) and in adding the "carry" in (3.18) is radix-p arithmetic and, from the point of view of the abbreviated representation, we proceed from left to right (rather than from right to left as in the case of decimals).

(3.20) Example. Multiply 2/3 by 1/6 in $\mathbf{Q}_5$. In Example 3.7 we find the 5-adic expansions for 2/3 and 5/6. It is easily verified that the 5-adic expansion for 1/6 is obtained from the 5-adic expansion for 5/6 by a single shift of the p-adic point. Hence,

$$\frac{1}{6} = .14040404\ldots \qquad (p = 5)$$

If we multiply these expansions, we obtain

```
 .4131313131313...
 .1404040404040...
 -----------------
  4131313131313...
   123131313131...
     1231313131...
       12313131...
         123131...
           1231...
             12...
 -----------------
 .4201243201243...
```

As a check we observe that (2/3)(1/6) = 1/9 and

$$\frac{1}{9} = .4\,201243\,201243\ldots \qquad (p = 5).$$

Division

Using an argument similar to that used for multiplication, there is no loss in generality in restricting our discussion of p-adic division to the division of p-adic units. Consequently, consider the p-adic units

$$(3.21) \qquad \begin{cases} \delta = d_0 + d_1 p + d_2 p^2 + \ldots \\ \beta = b_0 + b_1 p + b_2 p^2 + \ldots \end{cases}$$

with $d_0 b_0 \neq 0$. The quotient $\alpha = \delta/\beta$ can be written

$$(3.22) \qquad \alpha = \frac{d_0 + d_1 p + d_2 p^2 + \ldots}{b_0 + b_1 p + b_2 p^2 + \ldots} = a_0 + a_1 p + a_2 p^2 + \cdots$$

where we assume that $a_0, a_1, a_2, \ldots$ are digits.

To compute a_0 we observe that $\delta = \alpha\beta$ and if we form $\alpha\beta$, as in (3.15), we are led to the equations (3.16). These equations give us the coefficients $c_0, c_1, \ldots$ which are not digits and we must use (3.17) and (3.18) to get the digits $d_0, d_1, \ldots$ in (3.19). Observe that (3.16) has the equivalent matrix form

$$(3.23) \qquad \begin{bmatrix} b_0 & & & & \\ b_1 & b_0 & & & \\ b_2 & b_1 & b_0 & & \\ \cdot & \cdot & \cdot & \cdot & \cdot \\ b_k & b_{k-1} & b_{k-2} & \cdots & b_0 \\ \cdot & \cdot & \cdot & \cdot & \cdot \end{bmatrix} \begin{bmatrix} a_0 \\ a_1 \\ a_2 \\ \vdots \\ a_k \\ \vdots \end{bmatrix} = \begin{bmatrix} c_0 \\ c_1 \\ c_2 \\ \vdots \\ c_k \\ \vdots \end{bmatrix}$$

In both (3.16) and (3.23) we have $a_0 b_0 = c_0$. However, from (3.17), we have $c_0 = d_0 + t_1 p$. Hence,

$$(3.24) \qquad a_0 b_0 = d_0 + t_1 p,$$

from which we obtain $|a_0 b_0|_p = d_0$. Therefore,

$$(3.25) \qquad a_0 = |d_0 b_0^{-1}|_p \ .$$

In other words, to get the first digit of the quotient we form $b_0^{-1}(p)$, multiply this by d_0, and reduce the result modulo p.

It turns out that this is the clue for obtaining each digit of the expansion for α in (3.22). At each stage of the standard "long division" procedure we multiply $b_0^{-1}(p)$ by the first digit of the partial remainder and reduce the result modulo p. (At the first step, the dividend is considered to be the partial remainder.).

3.26 Example. Divide 2/3 by 1/12 in $\mathbf{Q}_5$. We have

$$\frac{2}{3} = .4131313... \qquad (p = 5)$$

and

$$\frac{1}{12} = .34\,24242... \qquad (p = 5).$$

In the "long division" which follows we do our subtraction at each step by complementing the subtrahend and adding.

The first digit of the divisor is $b_0 = 3$ and its multiplicative inverse is

$$b_0^{-1}(p) = 3^{-1}(5)$$
$$= 2 \ .$$

The first digit of the partial remainder (in the first step this is the dividend) is $d_0 = 4$ which gives us

$$a_0 = |4 \cdot 2|_5$$
$$= 3 \; .$$

Thus, the first step of the long division procedure is

```
                  .3
.3424242... ) .4131313...
               1111111...
               342424...
```

and the new partial remainder is 342424...

To get the second digit in the quotient we multiply $b_0^{-1}(p)$ by the first digit of the partial remainder and reduce the result modulo p. Hence,

$$a_1 = |3 \cdot 2|_5$$
$$= 1.$$

Thus, the second step of the long division procedure gives us

```
                 .31
.3424242... ) .4131313...
               1111111...
                342424...
                202020...
                 00000...
```

In this particular example we have produced a partial remainder which is zero. Hence, we terminate the expansion at this point because all remaining digits of the quotient are zero. In general, this does not happen and we continue until the period in the expansion has been exhibited.

As a check we observe that $2/3 \div 1/12 = 8$

and

$$8 = .3100000... \qquad (p = 5).$$

3.27 Remark. We point out that division of p-adic numbers is deterministic and not subject to trial and error as is the case with the division of decimals. This is due to the fact that, once we have $b_0^{-1}(p)$, where b_0 is the first digit of the divisor, the algorithm for obtaining each digit of the quotient is very specific; we multiply the first digit of each partial remainder by $b_0^{-1}(p)$ and reduce the result modulo p.

4. A Finite-segment p-adic Number System

A finite number system based on the system of p-adic numbers described in previous sections has been proposed recently by Krishnamurthy, Rao, and Subramanian [1975a and 1975b] and by Alparslan [1975]. In this finite number system each rational number in the set

$$(4.1) \qquad \mathbf{S}_N = \left\{ \alpha = \frac{a}{b} : 0 \leq |a| \leq N \text{ and } 0 < |b| \leq N \right\},$$

(where N is the positive integer defined in Theorem 4.7) is assigned a unique coded representation called its Hensel code. Arithmetic operations (addition, subtraction, multiplication, and division) on pairs of rational numbers in $\mathbf{S}_N$ can be replaced by corresponding arithmetic operations on their Hensel codes.

Hensel codes

The Hensel code for a rational number α is simply a finite segment of its p-adic expansion. We use the notation $H(p, r, \alpha)$, where p is a prime and r is the positive integer which specifies the number of digits of the p-adic expansion which we retain for the Hensel code. For example, since

$$(4.2) \qquad \frac{2}{3} = .4131313... \qquad (p = 5)$$

the Hensel code for $\alpha = 2/3$, when $p = 5$ and $r = 4$, is

$$(4.3) \qquad H(5, 4, 2/3) = .4131$$

In order to discuss the set $\mathbf{S}_N$ in (4.1) in more detail we need to introduce the following material.

4.4 Definition. A Farey sequence[†] of order N is the ascending sequence of all reduced fractions in [0, 1] whose denominators are not greater than N.

For example, if N = 5, we have the Farey sequence

$$(4.5) \qquad F_N = \frac{0}{1}, \frac{1}{5}, \frac{1}{4}, \frac{1}{3}, \frac{2}{5}, \frac{1}{2}, \frac{3}{5}, \frac{2}{3}, \frac{3}{4}, \frac{4}{5}, \frac{1}{1} \; .$$

4.6 Definition. An order-N Farey fraction is a fraction $\alpha = a/b$ for which $0 \leq |a| \leq N$ and $0 < |b| \leq N$.

Obviously, $\mathbf{S}_N$ in (4.1) is the set of all order-N Farey fractions. It is clear that $\mathbf{S}_N$ contains F_N as a proper subset.

We now state the existence and uniqueness theorem for Hensel codes. This theorem is fundamental in the development of finite-segment p-adic arithmetic.

4.7 Theorem. Let p be a prime and let r be a positive integer. Define N to be the largest positive integer which satisfies the inequality

$$N \leq \left[\frac{p^r - 1}{2}\right]^{\frac{1}{2}}.$$

[†]Named for the British geologist J. Farey (1766-1826).

Then every order-N Farey fraction $\alpha = a/b$ can be represented uniquely by an r-digit ordered sequence (its Hensel code), where each digit is an integer lying in the interval $[0, p-1]$.

Proof See Rao [1975].

As the theorem states, the r-digit representation is the Hensel code fo α and we denote it by writing

$$(4.8) \qquad H(p, r, \alpha) = a_n a_{n+1} \cdots a_{-2} a_{-1} \,.\, a_0 a_1 a_2 \cdots a_t$$

where $r = n+t+1$. In (4.8) we show n digits to the left of the p-adic point and $t+1$ digits to the right. Since the Hensel code is merely a finite segment of the p-adic expansion for α, we keep the p-adic point in the same position as in the p-adic expansion. (See Remark 2.18).

Residue equivalent of the Hensel code

We now describe an algorithm for mapping a rational number α onto its Hensel code $H(p, r, \alpha)$ which does not involve finding the (infinite) p-adic expansion and truncating it to r digits. It is based on the following result.

4.9 Theorem. Suppose $\alpha = a/b$, where $a/b = (c/d)p^n$, with

$$(c, d) = (c, p) = (d, p) = 1.$$

Let the Hensel code for c/d be

$$H(p, r, c/d) = .a_0 a_1 \cdots a_{r-1}$$

Then $a_{r-1}\ldots a_1a_0$ is the radix-p representationfor the integer $|c\cdot d^{-1}|_{p^r}$; in other words,

$$|c\cdot d^{-1}|_{p^r} = a_0 + a_1p + a_2p^2 + \ldots + a_{r-1}p^{r-1}.$$

Proof See Rao [1975].

Note that this theorem demonstrates the relationship which exists between the finite-segment p-adic representation (the Hensel code) and the residue representation of a rational number. Consider the following three cases:

Case I $\quad n = 0$

In this case $a = c$ and $b = d$ in Theorem 4.9. The first step in finding $H(p, r, \alpha)$ is to compute the integer $|c\cdot d^{-1}|_{p^r}$. The second step is to express this (decimal) integer as a radix-p integer. The third step is to reverse the order of the digits.

For example, let $\alpha = 2/3$, $p = 5$, and $r = 4$. Then $c = 2$, $d = 3$, $n = 0$, and $p^r = 625$. Hence,

$$(4.10) \qquad \left|\frac{2}{3}\right|_{625} = |2\cdot 3^{-1}|_{625}$$

$$= |2\cdot 417|_{625}$$

$$= 209.$$

Now

$$(4.11) \qquad 209_{ten} = 1314_{five} \cdot$$

If we write the radix-5 digits in reverse order, we obtain

$$(4.12) \qquad H(5, 4, 2/3) = .4131$$

and this agrees with (4.3).

Case II $\qquad n < 0$

In this case $\alpha = (c/d)p^{-m}$, where m is a positive integer. To find $H(p, r, \alpha)$ we find $H(p, r, c/d)$, using the three steps in Case I, and then shift the p-adic point m places to the right.

For example, let $\alpha = 2/15$, $p = 5$, and $r = 4$. We write $\alpha = (2/3)5^{-1}$ which gives us $c = 2$, $d = 3$, $m = 1$, and $p^r = 625$. Since we already have $H(5, 4, 2/3)$ in (4.12), we merely shift the p-adic point one place to the right to obtain

$$(4.13) \qquad H(5, 4, 2/15) = 4.131$$

Case III $\qquad n > 0$

In this case $\alpha = (c/d)p^k$, where k is a positive integer. To find $H(p, r, \alpha)$ we find $H(p, r, c/d)$ using the three steps in Case I, and then shift the p-adic point k places to the left.

For example, let $\alpha = 10/3$, $p = 5$, and $r = 4$. We write $\alpha = (2/3)5$ which gives us $c = 2$, $d = 3$, $k = 1$, and $p^r = 625$. Since we already have $H(5, 4, 2/3)$ in (4.12), we merely shift the p-adic one place to the left to obtain

$$(4.14) \qquad H(5, 4, 10/3) = .0413$$

Notice that, since $r = 4$, we retain only four digits (including the zero) and the rightmost digit of $H(5, 4, 2/3)$ is discarded.

Hensel codes for negative rational numbers

In Theorem 2.19 we stated the relationship that exists between the (infinite) p-adic expansions for α and for $-\alpha$. In Example 2.20 we observed that

$$(4.15)\qquad \begin{cases} \frac{1}{3} = .2313131\ldots & (p = 5) \\ -\frac{1}{3} = .3131313\ldots & (p = 5). \end{cases}$$

Consequently, the corresponding Hensel codes (for $p = 5$ and $r = 4$) are

$$(4.16)\qquad \begin{cases} H(5, 4, 1/3) = .2313 \\ H(5, 4, -1/3) = .3131 \end{cases}$$

Notice that the leftmost (non-zero) digit of the Hensel code for a positive rational number is complemented with respect to p and subsequent digits are complemented with respect to $p-1$. Thus, from (4.12), (4.13), and (4.14) we can write

$$(4.17)\qquad \begin{cases} H(5, 4, -2/3) = .1313 \\ H(5, 4, -2/15) = 1.313 \\ H(5, 4, -10/3) = .0131 \end{cases}$$

4.18 Remark. It should be pointed out that the algorithm we used in obtaining (4.12), (4.13), and (4.14) can be used for negative fractions as well as for positive fractions. It can also be used in those cases for which $(a,b) \neq 1$ and $(a,p) \neq 1$.

For example, let $\alpha = -2/4$. Then, for $p = 5$ and $r = 4$,

$$\begin{aligned} |-2/4|_{625} &= |(-2)\cdot 4^{-1}|_{625} \\ &= |(-2)\cdot 469|_{625} \\ &= 312. \end{aligned}$$

Since the radix-5 equivalent of this integer is

$$312_{ten} = 2222_{five} ,$$

we have the Hensel code

$$H(5,\ 4,\ -2/4) = .2222$$

It is easy to verify that $-1/2$, $-3/6$, $-4/8$, etc., all have this same Hensel code. We point out that

$$H(5,\ 4,\ 1/2) = .3222$$

and that .2222 and .3222 are complements of each other.

Similarly, if $\alpha = 10/3$ (the example we used in Case III, above), then

$$|10/3|_{625} = |10 \cdot 3^{-1}|_{625}$$

$$= |10 \cdot 417|_{625}$$

$$= 420.$$

Since the radix-5 equivalent of this integer is

$$420_{ten} = 3140_{five}$$

we obtain the Hensel code

$$H(5, 4, 10/3) = .0413$$

and this agrees with (4.14) above.

Floating-point Hensel codes

In discussing (4.14), in Case III above, we made the observation that H(5, 4, 10/3) and H(5, 4, 2/3) are related by a shift of the p-adic point and that one of the "significant digits" in the Hensel code for 2/3 is discarded when we generate the Hensel code for 10/3. This can be avoided if we introduce the concept of a normalized floating-point Hensel code.

4.19 Definition. Let $\alpha = a/b$ with $a/b = (c/d)p^n$ and

$$(c,d) = (c,p) = (d,p) = 1.$$

Then

$$\hat{H}(p, r, \alpha) = (m_\alpha, e_\alpha),$$

with

$$m_\alpha = H(p, r, c/d)$$

and

$$e_\alpha = n,$$

is the normalized floating-point Hensel code for α. We call m_α the <u>mantiss</u> and e_α the <u>exponent</u>.

4.20 <u>Examples</u>. From (4.12), (4.13), and (4.14) we obtain

$$\hat{H}(5, 4, 2/3) = (.4131, 0)$$

$$\hat{H}(5, 4, 2/15) = (.4131, -1)$$

$$\hat{H}(5, 4, 10/3) = (.4131, 1).$$

and from (4.17) we obtain

$$\hat{H}(5, 4, -2/3) = (.1313, 0)$$

$$\hat{H}(5, 4, -2/15) = (.1313, -1)$$

$$\hat{H}(5, 4, -10/3) = (.1313, 1)$$

Notice that the digit in H(5, 4, 2/3) which was discarded when we formed H(5, 4, 10/3) in (4.14) has been recovered when we form the floating-point Hensel code for 10/3. This will be significant when we examine arithmetic using Hensel codes as operands.

4.21 Remark. In the normalized floating-point Hensel codes notice that the mantissa m_α is the ordinary Hensel code for c/d (in Definition 4.19). Also notice that the p-adic point in m_α is at the left of the string of r digits and that the leftmost digit (the one next to the p-adic point) is non-zero. In other words, c/d is a p-adic unit (see Definition 3.11) and m_α is a finite segment of the infinite p-adic expansion for that p-adic unit. Thus, the mantissas in the normalized floating-point Hensel codes play the same role in this finite system as the p-adic units play in the infinite system.

4.22 Table The Ordinary Hensel Codes[†] H(5, 4, a/b).

b \ a	1	2	3	4	5	6	7	
1	.1000	.2000	.3000	.4000	.0100	.1100	.2100	.3
2	.3222	.1000	.4222	.2000	.0322	.3000	.1322	.4
3	.2313	.4131	.1000	.3313	.0231	.2000	.4313	.1
4	.4333	.3222	.2111	.1000	.0433	.4222	.3111	.2
5	1.000	2.000	3.000	4.000	.1000	1.100	2.100	3.1
6	.1404	.2313	.3222	.4131	.0140	.1000	.2404	.3
7	.3302	.1214	.4021	.2423	.0330	.3142	.1000	.4
8	.2414	.4333	.1303	.3222	.0241	.2111	.4030	.1
9	.4201	.3012	.2313	.1124	.0420	.4131	.3432	.2
10	3.222	1.000	4.222	2.000	.3222	3.000	1.322	4.0
11	.1332	.2120	.3403	.4240	.0133	.1411	.2204	.3
12	.3424	.1404	.4333	.2313	.0342	.3222	.1202	.4
13	.2034	.4014	.1143	.3123	.0203	.2232	.4212	.1
14	.4101	.3302	.2013	.1214	.0410	.4021	.3222	.2
15	2.313	4.131	1.000	3.313	.2313	2.000	4.313	1.2
16	.1234	.2414	.3104	.4333	.0123	.1303	.2042	.3
17	.3043	.1132	.4121	.2210	.0304	.3342	.1431	.4

b \ a	9	10	11	12	13	14	15	16
1	.4100	.0200	.1200	.2200	.3200	.4200	.0300	.1300
2	.2322	.0100	.3322	.1100	.4322	.2100	.0422	.3100
3	.3000	.0413	.2231	.4000	.1413	.3231	.0100	.2413
4	.1433	.0322	.4111	.3000	.2433	.1322	.0211	.4000
5	4.100	.2000	1.200	2.200	3.200	4.200	.3000	1.300
6	.4222	.0231	.1140	.2000	.3404	.4313	.0322	.1231
7	.2214	.0121	.3423	.1330	.4142	.2000	.0402	.3214
8	.3414	.0433	.2303	.4222	.1241	.3111	.0130	.2000
9	.1000	.0301	.4012	.3313	.2124	.1420	.0231	.4432
10	2.322	.1000	3.322	1.100	4.322	2.100	.4222	3.100
11	.4324	.0212	.1000	.2332	.3120	.4403	.0340	.1133
12	.2111	.0140	.3020	.1000	.4424	.2404	.0433	.3313
13	3321	.0401	.2430	.4410	.1000	.3034	.0114	.2143
14	.1134	.0330	.4431	.3142	.2343	.1000	.0201	.4302
15	3.000	.4131	2.231	4.000	1.413	3.231	.1000	2.413
16	.4402	.0241	.1421	.2111	.3340	.4030	.0310	.1000
17	.2024	.0113	.3102	.1240	.4234	.2323	.0412	.3401

[†]See Krishnamurthy, Rao, and Subramanian [1973a], page 70. Reproduced with permission.

Two rational numbers with the same Hensel code

Consider the two distinct rational numbers α and β and their canonical p-adic expansions

$$(4.23)\quad \begin{cases} \alpha = a_n p^n + a_{n+1} p^{n+1} + \cdots \\ \beta = b_n p^n + b_{n+1} p^{n+1} + \cdots \end{cases}$$

Even though $\alpha \neq \beta$ it is still possible for the r leading coefficients of α to be identical with the r leading coefficients of β, in which case $H(p, r, \alpha)$ will equal $H(p, r, \beta)$. The following theorem gives a characterization of this situation.

4.24 Theorem. Let $\alpha, \beta \in \mathbf{Q}$. Then $H(p, r, \alpha) = H(p, r, \beta)$ if and only if p^r divides $\alpha - \beta$.

Proof Let $H(p, r, \alpha) = H(p, r, \beta)$. Then

$$\alpha = a_n p^n + \cdots + a_{n+r-1} p^{n+r-1} + a_{n+r} p^{n+r} + \cdots$$

and

$$\beta = a_n p^n + \cdots + a_{n+r-1} p^{n+r-1} + b_{n+r} p^{n+r} + \cdots$$

where n can be positive, negative, or zero. If we form $\alpha - \beta$, we obtain the expansion

$$\begin{aligned} \alpha - \beta &= (a_{n+r} - b_{n+r}) p^{n+r} + (a_{n+r+1} - b_{n+r+1}) p^{n+r+1} + \cdots \\ &= p^r \left[(a_{n-r} - b_{n-r}) p^n + (a_{n+r+1} - b_{n+r+1}) p^{n+1} + \cdots \right] \end{aligned}$$

and p^r divides $\alpha - \beta$. We leave the proof of the converse to the reader.

Notice that if $\alpha = a/b$ and $\beta = g/h$, with $(b,p) = (h,p) = 1$, then α and β can be represented by the integers $|a \cdot b^{-1}|_{p^r}$ and $|g \cdot h^{-1}|_{p^r}$, respectively. In Remark 4.18 we pointed out that these two integers, when written in radix-p notation, give us the Hensel codes for α and β (when we reverse the order of the digits). Thus, it is easy to prove the followi result.

4.25 Theorem. Let $\alpha = a/b$ and $\beta = g/h$, with $(b,p) = (h,p) = 1$. Then $H(p, r, \alpha) = H(p, r, \beta)$ if and only if

$$|a \cdot b^{-1}|_{p^r} = |g \cdot h^{-1}|_{p^r},$$

that is, if and only if

$$a \cdot b^{-1} \equiv g \cdot h^{-1} \pmod{p^r}.$$

4.26 Corollary. Let $\alpha = a/b$ and $\beta = g/h$, with $(b,p) = (h,p) = 1$. Then $H(p, r, \alpha) = H(p, r, \beta)$ if and only if

$$|ah|_{p^r} = |bg|_{p^r},$$

that is, if and only if

$$ah \equiv bg \pmod{p^r}.$$

Proof In Theorem 4.25 multiply both sides of the equation and the congruenc by bh and simplify.

4.27 <u>Example.</u>† Let $p = 5$, $r = 4$, so that $p^r = 625$. Consider the two rational numbers $\alpha = 10/13$ and $\beta = -35/17$. Obviously, 625 divides $\alpha - \beta = 625/221$. Observe also that

$$\begin{aligned} 10 \cdot 13^{-1}(625) &= 10 \cdot 577 \\ &= 5770, \end{aligned}$$

and

$$\begin{aligned} -35 \cdot 17^{-1}(625) &= -35 \cdot 478 \\ &= -16730, \end{aligned}$$

and that

$$5770 \equiv -16730 \quad (\text{mod } 625).$$

Finally, observe that

$$10 \cdot 17 \equiv 13(-35) \quad (\text{mod } 625).$$

Consequently, from either Theorem 4.24, Theorem 4.25, or Corollary 4.26 we have

$$H(5, 4, 10/13) = H(5, 4, -35/17).$$

Notice that

$$\begin{aligned} |10 \cdot 13^{-1}|_{625} &= |-35 \cdot 17^{-1}|_{625} \\ &= 145 \end{aligned}$$

and if we convert this decimal number to its radix-5 representation, we obtain

$$145_{ten} = 1040_{five}.$$

Consequently,

$$\begin{aligned} H(5, 4, 10/13) &= H(5, 4, -35/17) \\ &= .0401 \end{aligned}$$

---

†This example is due to Ruth Ann Lewis.

and this value agrees with $H(5, 4, 10/13)$ in Table 4.22.

Obviously, any rational number a/b for which

$$|a \cdot b^{-1}|_{625} = 145 \tag{4.28}$$

has the same Hensel code as $10/13$. However, among such rational numbers, o 10/13 is an order-17 Farey fraction, that is, only $10/13$ lies in $\mathbf{S}_{17}$. T example illustrates the general principle that corresponding to a given Hens code there may be many rational numbers but only one will lie in $\mathbf{S}_N$. (We a assuming that this unique element in $\mathbf{S}_N$ is reduced because we observed in Remark 4.18 that $-1/2$, $-3/6$, $-4/8$, etc., all have the same Hensel code.)

4.29 Remark. It should be emphasized that the Hensel codes referred to in Theorem 4.24 and Theorem 4.25 are ordinary Hensel codes, not normalized floa point Hensel codes. For example,

$$\hat{H}(5, 4, 10/13) = (.4014, 1)$$

whereas

$$\hat{H}(5, 4, -35/17) = (.4013, 1)$$

and the mantissas differ in the fourth digit.

It is clear that agreement between floating-point Hensel codes implies agreement between ordinary Hensel codes but the converse is not necessarily

5. Arithmetic Operations On Hensel Codes

In Section 3 we discussed arithmetic operations in the field of p-adic numbers $\mathbb{Q}_p$. We illustrated addition, subtraction, multiplication, and division using (infinite) p-adic expansions as operands. Since the Hensel codes are merely finite segments of the p-adic expansions it is not surprising to discover that we can do arithmetic using the Hensel codes as operands. In addition, it is not surprising to discover that the rules for the arithmetic of Hensel codes are essentially the same as the rules for arithmetic in $\mathbb{Q}_p$.

Throughout this section we shall assume that the operands are normalized floating-point Hensel codes with $p = 5$ and $r = 4$ unless it is stated otherwise.

Addition

Consider the following numerical example:

$$\frac{2}{3} + \frac{1}{5} = \frac{13}{15} . \tag{5.1}$$

The Hensel codes

$$\begin{cases} \hat{H}(5, 4, 2/3) = (.4131, 0) \\ \\ \hat{H}(5, 4, 1/5) = (.1000, -1) \end{cases} \tag{5.2}$$

can be used to carry out this addition. We simply line up the p-adic points and do radix-5 arithmetic proceeding from left to right. Since $r = 4$, we have

$$\begin{array}{r} .4131 \\ \underline{1.000} \\ 1.413 \end{array}$$

Hence, the sum we seek has .1413 as its mantissa and -1 as its exponent, that is,

$$\hat{H}(5,\ 4,\ \alpha) = (.1413,\ -1)\ . \tag{5.3}$$

Since

$$\hat{H}(5,\ 4,\ 13/15) = (.1413,\ -1) \tag{5.4}$$

and since the Hensel code is unique, we have

$$\alpha = \frac{13}{15}\ , \tag{5.5}$$

and this is the correct answer.

Subtraction

Just as we did in $\mathbf{Q}_p$, we treat subtraction as "complemented addition" in the sense that the subtrahend is complemented and added to the minuend. example, to carry out the computation

$$\frac{2}{3} - \frac{1}{5} = \frac{7}{15} \tag{5.6}$$

using Hensel codes we need

$$\hat{H}(5,\ 4,\ -1/5) = (.4444,\ -1)\ . \tag{5.7}$$

Then, as before, we line up the p-adic points and do radix-5 addition procee from left to right. Since $r = 4$, we have

$$\begin{array}{r} .4131 \\ 4.444 \\ \hline 4.313 \end{array}$$

Hence, the answer we seek has .4313 as its mantissa and -1 as its exponen that is,

(5.8) $$\hat{H}(5, 4, \alpha) = (.4313, -1) .$$

Since

(5.9) $$\hat{H}(5, 4, 7/15) = (.4313, -1)$$

and since the Hensel code is unique, we have

(5.10) $$\alpha = \frac{7}{15} ,$$

and this is the correct answer.

Multiplication

Consider the following numerical example:

(5.11) $$\frac{1}{3} \cdot \frac{6}{5} = \frac{6}{15} .$$

We carry out this computation using the floating-point Hensel codes

(5.12) $$\begin{cases} \hat{H}(5, 4, 1/3) = (.2313, 0) \\ \hat{H}(5, 4, 6/5) = (.1100, -1) . \end{cases}$$

The algorithm involves multiplying the mantissas and adding the exponents. Hence, using radix-5 arithmetic (proceeding from left to right), we obtain

$$\begin{array}{r} .2313 \\ \underline{.1100} \\ 2313 \\ \underline{231} \\ .2000 \end{array}$$

as the mantissa and $0 + (-1) = -1$ as the exponent for the product, that is,

(5.13) $$\hat{H}(5, 4, \alpha) = (.2000, -1) .$$

Since

$$\hat{H}(5,\ 4,\ 6/15) = (.2000,\ -1)$$

and since the Hensel code is unique, we have

(5.15) $$\alpha = \frac{6}{15}\ ,$$

and this is the correct result.

Division

Consider the following numerical example:

(5.16) $$\frac{2}{15} \div \frac{4}{15} = \frac{1}{2}\ .$$

We carry out this computation using the floating-point Hensel codes

(5.17) $$\begin{cases} \hat{H}(5,\ 4,\ 2/15) = (.4131,\ -1) \\ \hat{H}(5,\ 4,\ 4/15) = (.3313,\ -1)\ . \end{cases}$$

The algorithm involves dividing the mantissas and subtracting the expon We divide the mantissas the way we divided p-adic units in Section 3. (See 3.26). We need the multiplicative inverse, modulo p, of the leftmost digit divisor. In this example, we divide .4131 by .3313 and so we need

(5.18) $$3^{-1}(5) = 2.$$

We proceed exactly as we did in Example 3.26 except that we terminate as soo we have r digits in the quotient. Hence,

```
         .3222
.3313 ) .4131
         1444
         ----
          131
          421
          ---
           13
           42
           --
            1
            4
            -
            0
```

and so the quotient we seek has .3222 for its mantissa. When we subtract exponents we get zero. Hence,

$$\hat{H}(5, 4, \alpha) = (.3222, 0) .$$

Since

$$\hat{H}(5, 4, 1/2) = (.3222, 0)$$

and since the Hensel code is unique, we have

$$\alpha = 1/2,$$

and this is the correct result.

6. Removing a Leading Zero From a Hensel Code

In Definition 4.19 we assumed that

$$\alpha = (c/d)p^n \tag{6.1}$$

with $(c,d) = (c,p) = (c,p) = 1$ and this enables us to write the normalized floating-point Hensel code in the form

$$\hat{H}(p,r,\alpha) = (m_\alpha, e_\alpha) , \tag{6.2}$$

where

$$m_\alpha = H(p,r,c/d) . \tag{6.3}$$

In other words, the mantissa in the floating-point Hensel code is the ordina Hensel code for c/d. In this case, of course, the leftmost digit of m_α is different from zero which means m_α can be used as the divisor in a divisio operation.

There are occasions following the operations of addition and/or subtrac when a Hensel code contains a zero in the leftmost digit position and yet th Hensel code is supposed to be the divisor in a division operation. Since th leftmost digit in the Hensel code of a divisor must have a multiplicative inverse modulo p , this creates a problem.

6.4 Example Suppose we wish to compute

$$x = \frac{a}{b + c}$$

with $b = 1/2$ and $c = 1/8$. If we use the Hensel codes

$$\hat{H}(5,4,1/2) = (.3222,\ 0)$$

and

$$\hat{H}(5,4,1/8) = (.2414,\ 0)$$

we obtain the mantissa

$$\begin{array}{r} .3222 \\ .2414 \\ \hline .0241 \end{array}$$

that is, the ordinary Hensel code for b + c = 5/8 is .0241.

To use b + c as the divisor in evaluating x we need the normalized floating-point Hensel code for 5/8 which is

$$\hat{H}(5,4,5/8) = (.2414,\ 1)\ .$$

Hence, the big question is how to obtain the mantissa .2414 when we only know .0241 .

To answer the question raised by Example 6.4 we need a procedure for mapping an ordinary Hensel code onto its rational equivalent u/v so that we can then map u/v onto its normalized floating-point Hensel code. There is such a procedure if we know an integer multiple of v , say mv . In this case,

(6.5) $$mu = \Big/ mv \left| u \cdot v^{-1} \right|_{p^r} \Big/_{p^r}\ ,$$

(which resembles (2.27) in Chapter 4) and

(6.6) $$u/v = mu/mv\ .$$

In Example 6.4 we produced the Hensel code .0241 as a result of an addition. If we reverse the order of the digits we obtain the radix-five integer 1420. Since

(6.7) $$1420_{five} = 235_{ten}$$

and since $p = 5$, $r = 4$ implies $p^r = 625$, we have

(6.8) $$| u \cdot v^{-1} |_{625} = 235 .$$

Because $b + c$ in Example 6.4 involved adding 1/2 to 1/8 we know that the least common denominator of the two rational numbers is 8. Hence, we can assume that $mv = 8$. If we use this value and the result displayed in (6.8), then (6.5) gives us

(6.9) $$\begin{aligned} mu &= / 8 \cdot 235 /_{625} \\ &= 5 . \end{aligned}$$

Thus, the rational equivalent of the Hensel code .0241 is

(6.10) $$u/v = 5/8 .$$

Since $5/8 = (1/8)5$ and since

(6.11) $$H(5,4,1/8) = .2414$$

the normalized floating-point Hensel code for 5/8 is

(6.12) $$\hat{H}(5,4,5/8) = (.2414, 1) .$$

7. Mapping a Hensel Code Onto Its Rational Equivalent

In the previous section we discussed a procedure for mapping a Hensel code onto its rational equivalent u/v under the assumption that we somehow knew an integer multiple of v. We also need a procedure when no multiple of v is known. A simple (but not very efficient) procedure makes use of the fact that the Hensel code of an integer is easily recognized.

It is shown in Krishnamurthy, Rao, and Subramanian [1975a] that if r is even and α is a positive integer, then the last $r/2$ digits in $H(p,r,\alpha)$ are all zero. Similarly, if α is a negative integer the last $r/2$ digits in $H(p,r,\alpha)$ are all $p-1$. For example, if $p = 5$ and $r = 8$,

$$H(5,8,199) = .44210000 \tag{7.1}$$

and

$$H(5,8,-199) = .10234444 \; . \tag{7.2}$$

When we have the Hensel code for a rational number u/v (not an integer) we can add it to itself v times and produce the Hensel code for the integer u because of the simple fact that

$$v(u/v) = u \; . \tag{7.3}$$

Obviously, the number of additions required to produce the recognizable Hensel code is v.

If $u < v$ it is more efficient to use this procedure with the Hensel code for v/u than with the Hensel code for u/v. In this case we add the Hensel code to itself u times to produce the recognizable Hensel code for the integer v.

The best procedure is to operate with both Hensel codes simultaneously and terminate after t additions where

$$t = \min(u,v) . \tag{7.4}$$

7.5 Remark Despite the many advantages described in this chapter, there are certain difficulties associated with finite-segment p-adic arithmetic which need further study. These include

(i) the detection of the sign and magnitude of a/b when only the Hens code is known.

(ii) the detection of overflow when either the numerator or the denomina of a computational result exceeds N, and

(iii) the need for more efficient algorithms for mapping Hensel codes ont their rational equivalents.

REFERENCES

Adegbeyeni, E.O. [1977], "Finite-field computation technique for exact solution of systems of linear equations and interval linear programming problems", Int. J. Syst., 9, 1181-1192.

Agnew, J. [1972], Explorations in Number Theory, Brooks/Cole, Monterey, Calif.

Alparslan, E. [1975], "Finite p-adic number systems with possible applications", Ph.D. dissertation, Department of Electrical Engineering, University of Maryland.

Bachman, G. [1964], Introduction to P-Adic Numbers and Valuation Theory, Academic Press, New York.

Bauer, F. L. [1963], "Optimally scaled matrices", Numer. Math., 5, 73-87.

Beiser, P. S. [1979], An examination of finite-segment p-adic number systems as an alternative methodology for performing exact arithmetic", M.S. thesis, School of Engineering and Applied Science, University of Virginia, Charlottesville.

Borevich, Z. I. and Shafarevich, I. R. [1966], Number Theory, Academic Press, New York.

Businger, P.A. [1968], "Matrices which can be optimally scaled", Numer. Math., 12, 346-348.

Farinmade, J. A. [1976], "Fast, parallel, exact matrix computations using p-adic arithmetic", M.S. Thesis, Department of Computer Science, University of Lagos, Nigeria.

Forsythe, G. E. and Moler, C. B. [1967], Computer Solution of Linear Algebraic Systems, Prentice-Hall, Englewood Cliffs, N.J.

Forsythe, G. E. [1969], "Solving a quadratic equation on a computer", An essay in The Mathematical Sciences, A Collection of Essays, edited by the National Research Council's COSRIMS, M.I.T. Press, Cambridge, Mass.

Forsythe, G. E., Malcolm, M. A., and Moler, C. B. [1977], Computer Methods for Mathematical Computations, Prentice-Hall, Englewood Cliffs, N.J.

Givens, J. W. [1954], "Numerical computation of the characteristic values of a real symmetric matrix", Oak Ridge National Laboratory Report, ORNL-1574.

Gregory, R. T. and Karney, D. L. [1978], A Collection of Matrices for Testing Computational Algorithms, Krieger Pub. Co., Huntington, N.Y.

Gregory, R. T. [1978], "The use of finite-segment p-adic arithmetic for exact computations", BIT, 18, 282-300.

Hardy, G. H. and Wright, E. M. [1960], An Introduction to the Theory of Numbers, Clarendon Press, Oxford.

Hehner, E. C. R. and Horspool, R. N. S. [1978], "Exact arithmetic using a variable-length p-adic representation", Proceedings of the 4th IEEE Symposium on Computer Arithmetic, 10-14.

Hehner, E. C. R. and Horspool, R. N. S. [1979], "A new representation of the rational numbers for fast easy arithmetic", SIAM J. Comput., 8, 124-134.

Hensel, K. [1908], Theorie der Algebraischen Zahlen, Teubner, Leipzig-Stuttgart.

Howell, J. A. and Gregory, R. T. [1970], "Solving linear equations using residue arithmetic--Algorithm II", BIT, 10, 23-37.

Kline, M. [1972], Mathematical Thought from Ancient to Modern Times, Oxford University Press, New York.

Knuth, D. E. [1968], The Art of Computer Programming: Volume I Fundamental Algorithms, Addison Wesley, Reading, Mass.

Koblitz, N. [1977], P-Adic Numbers, P-Adic Analysis, and Zeta Functions, Springer-Verlag, New York.

Krishnamurthy, E. V., Rao, T. M., and Subramanian, K. [1975a], "Finite segment p-adic number systems with applications to exact computation", Proc. Indian Acad. Sci., 81A, 58-79.

Krishnamurthy, E. V., Rao, T. M., and Subramanian, K. [1975b], "P-adic arithmetic procedures for exact matrix computations", Proc. Indian Acad. Sci., 82A, 165-175.

Krishnamurthy, E. V. [1977], "Matrix processors using p-adic arithmetic for exact linear computations", IEEE Transactions on Computers, C-26, 633-639.

Krishnamurthy, E. V. [1978] "Exact inversion of a rational polynomial matrix using finite field transforms", SIAM J. Appl. Math., 35, 453-464.

Kunz, K. S. [1957], Numerical Analysis, McGraw-Hill, New York.

Lewis, Ruth Ann [1979], "P-adic number systems for error-free computation", Ph.D. dissertation, Department of Mathematics, University of Tennessee, Knoxville.

MacDuffee, C. C. [1938], "The p-adic numbers of Hensel", Amer. Math. Monthly, 45, 500-508.

Mahler, K. [1973], Introduction to P-Adic Numbers and Their Functions, Cambridge Univ. Press.

McCoy, N. H. [1948], Rings and Ideals, Carus Mathematical Monograph No. 8, The Mathematical Association of America, Washington, D.C.

McKeeman, W. M. [1962], "Algorithm 135, Crout with equilibration and interation", Comm. Assoc. Comput. Mach., 5, 553-555.

Rao, T. M. [1975], "Finite field computational techniques for exact solution of numerical problems", Ph.D. dissertation, Department of Applied Mathematics, Indian Institute of Science, Bangalore, India.

Rao, T. M. [1978], "Error-free computation of characteristic polynomial of a matrix", Comp. and Math. with Appls., 4, 61-65.

Richards, R. K. [1955], Arithmetic Operations in Digital Computers, D. Van Nostrand, Princeton, N.J.

Stewart, G. W. [1973], Introduction to Matrix Computations, Academic Press, New York.

Szabó, N. S. and Tanaka, R. I. [1967], Residue Arithmetic and Its Applications to Computer Technology, McGraw-Hill, New York.

Varga, R. S. [1962], Matrix Iterative Analysis, Prentice-Hall, Englewood Cliffs, N.J.

von Neumann, J. and Goldstine, H. H. [1947], "Numerical inverting of matrices of high order", Bull. Amer. Math. Soc. 53, 1021-1099.

Wilkinson, J. H. [1961], "Error analysis of direct methods of matrix inversion", J. Assoc. Comput. Mach., 8, 281-330.

Wilkinson, J. H. [1963], Rounding Errors in Algebraic Processes, Prentice-Hall, Englewood Cliffs, N.J.

Wilkinson, J. H. [1965], The Algebraic Eigenvalue Problem, Clarendon Press, Oxford.

Young, D. M. and Gregory, R. T. [1972], A Survey of Numerical Mathematics, Vol. I, Addison Wesley, Reading, Mass.

Young, D. M. and Gregory, R. T. [1973], A Survey of Numerical Mathematics, Vol. II, Addison Wesley, Reading, Mass.

INDEX